Finite-element-model Updating Using Computional Intelligence Techniques

T. Marwala

Finite-element-model Updating Using Computional Intelligence Techniques

Applications to Structural Dynamics

Springer

Prof. Tshilidzi Marwala
University of Johannesburg
Faculty of Engineering and the Built Environment
Cnr Kingsway and University Road
Auckland Park 2092
South Africa
tmarwala@gmail.com

ISBN 978-1-4471-5716-8 ISBN 978-1-84996-323-7 (eBook)
DOI 10.1007/ 978-1-84996-323-7
Springer London Dordrecht Heidelberg New York

British Library Cataloguing in Publication Data
A catalogue record for this book is available from the British Library

Softcover re-print of the Hardcover 1st edition 2010

Cover design: eStudioCalamar, Figueres/Berlin

Printed on acid-free paper

Springer is part of Springer Science+Business Media (www.springer.com)

Foreword

Finite-element modeling in this book is viewed as the mathematical and numerical process through which a physical structure is translated into a mathematical model and from that mathematical model a numerical procedure is used to estimate such dynamic characteristics as mode shapes and natural frequencies. Finite-element updating is a process through which such models are tuned to better reflect the measured data.

In this book, the Nelder–Mead simplex and Broyden–Fletcher–Goldfarb–Shanno (BFGS) optimization methods are introduced, applied and compared for finite-element-model updating. The use of reduction and expansion methods to equate measured modal data to finite-element systems matrices is also investigated. Furthermore, genetic algorithms are introduced and applied to finite-element-model updating.

The particle-swarm optimization method is also introduced and applied for finite-element-model updating and the results are compared to those obtained from the genetic algorithm. Furthermore, simulated annealing is also introduced and applied to finite-element-model updating and the results are compared to those from particle-swarm optimization. To deal with the issue of computational efficiency, a response-surface method that combines the multi-layer perceptron and particle-swarm optimization is introduced and applied to finite-element-model updating. The results are compared to those from genetic algorithm, particle swarm optimization and simulated annealing.

To exploit the combined advantages of different optimization methods, a hybrid optimization method is introduced that combines particle-swarm optimization, with the Nelder–Mead simplex method and it is applied to finite-element-model updating. The results are compared to those from when genetic algorithm, particle-swarm optimization and simulated annealing are used individually.

Furthermore, a multi-objective optimization method that uses both modal properties data and frequency-domain data is introduced for finite-element-model updating. In addition, the multi-layer perceptron network is used for finite-element-model updating.

To bring the finite-element-model updating procedure onto firm statistical grounds, a Bayesian approach is applied. To illustrate the use of finite-element updating, an application of this procedure for damage detection in structures is conducted. Finally, the book concludes with key recommendations and outstanding issues for further development.

May 2010 *Snehashish Chakraverty PhD*
National Institute of Technology-Rourkela
Orissa, India

Preface

Finite-element-model Updating Using Computational Intelligence Techniques introduces the concepts of computational intelligence for finite-element-model updating. Finite-element modeling is a subject that has received acceptance and has applications in various disciplines of engineering including aerospace, civil, mechanical and electrical engineering. These finite-element models, however, do not necessarily predict the measured data sufficiently accurately. Because of this, there is a need for these models to be updated to better reflect the measured data.

This book introduces computational intelligence techniques to update finite-element models. The computational intelligence methods used for finite-element-model updating include neural networks, genetic algorithms, particle-swarm optimization, simulated annealing, response-surface methods, hybrid methods and Bayesian methods. Applications to engineering problems are considered especially for updating of finite-element models and its application to damage detection.

This book makes an interesting read and it will open up new avenues in the use of computational intelligence techniques to the problem of finite-element-model updating.

May 2010 *Tshilidzi Marwala, PhD*
University of Johannesburg, Johannesburg

Acknowledgements

I would like to thank the following institutions for contributing towards the writing of this book: University of Cambridge, University of Pretoria and University of Johannesburg.

I also would like to thank my following former and present graduate students for their assistance in developing this manuscript: Ishmael Msiza, Lesedi Masisi, and Linda Mthembu. In particular, I thank Dr. Ian Kennedy for carefully reviewing this book.

I dedicate this book to the schools that gave me the foundation to always seek excellence in everything I do and these are: Mbilwi Secondary School, Case Western Reserve University, University of Pretoria, University of Cambridge (St. John' College) and Imperial College (London).

I also thank my supervisors who played pivotal roles in my education and these are: Professor P.S. Heyns of the University of Pretoria, Dr. H.E.M. Hunt of the University of Cambridge and Professor Philippe de Wilde of Herriot-Watt University.

This book is dedicated to the following people: Dr. Jabulile Vuyiswa Manana, Mr. Nhlonipho Khathutshelo Marwala, Mrs Reginah Marwala and Mr. Shavhani Marwala.

Contents

Chapter 1

Introduction to Finite-element-model Updating

Abstract. This chapter introduces finite-element-model updating. Direct and iterative updating procedures are explained. Some basic features on finite-element modeling are elucidated. Essential elements on vibration testing and analysis are explained and these include the domains in which data can be represented. These domains are in the modal, frequency and time–frequency spaces. Finite-element-model updating techniques are then reviewed and these can be broadly categorized into: matrix update methods, sensitivity-based techniques, iterative optimization procedures, Bayesian methods and computational intelligence techniques. Computational intelligence technques, which are the subject of this book, are then reviewed in detail.

Keywords: finite-element-model updating, direct method, iterative method, frequency domain, modal domain, computational intelligence

1.1 Introduction

The development of modern computers capable of processing large matrices has led to the construction of many large and intricate numerical models. One of these numerical models is the finite-element model. The first application of finite-element techniques was in solving complex elasticity and structural analysis problems in aeronautical and civil engineering. Finite-element modeling was first developed by Hrennikoff (1941) as well as Courant and Robbins (1941). Courant used the Ritz methods as well as variational calculus to solve vibration problems (Hastings *et al.*, 1985). While the techniques used by these founders are very different from current approaches, some important characteristics are still shared. These differences include mesh discretization into elements (Babuska *et al.*, 2004).

The Cooley–Turkey algorithm and related methods, which are used to obtain Fourier transformations has led to the development of sophisticated methods in vibration and experimental modal analysis (see Appendices A and B). However, the finite-element model usually gives results that are not the same as the results given by an experiment. The reasons for the discrepancy between finite-element-model data and measured data include (Friswell and Mottershead, 1995):

- model structure errors, which may result from the difficulty of modeling damping, joints, welds and edges;
- model order errors, which may result from the difficulty in modeling nonlinearity;
- model parameter errors, which result in difficulty in identifying the correct material properties; and
- errors in measurements.

In this book, we assume that the measurements are correct and therefore the finite-element model must be updated to match the measured data. Furthermore, this book assumes that the difficulty in modeling joints and other complicated boundary conditions can be compensated for by adjusting the material properties of the relevant elements. In addition, it is assumed that the finite-element models are linear and that damping is low enough not to require complex attention. Due to this inconsistency between measured and finite-element data, computational methods have been developed to update the finite-element model so that it can closely predict measured results (Mottershead and Friswell, 1993; Friswell and Mottershead, 1995). Techniques developed to update the finite-element model fall into two categories: direct and iterative.

Direct methods update the finite-element model without any regard to changes in physical parameters. For this reason, direct methods tend to give models that represent the measured parameters without any regard to the structure that is being analyzed. This results in mass and stiffness matrices that have little physical meaning and cannot be related to physical changes in the finite-elements of the original model. Furthermore, the connectivity of the nodes is not ensured and, generally, the matrices are fully populated and not sparse.

When using *iterative methods,* physical parameters are updated until the finite-element model reproduces the measured data to a sufficient degree of accuracy. Because of this nature of iterative methods, they give finite-element models that ensure the connectivity of nodes, and have mass and stiffness matrices that have physical meaning. With the aim of using the proposed updating method on damage detection, iterative techniques are adopted in this book.

1.2 Finite-element Modeling

Many disciplines such as aerospace, civil, and mechanical engineering normally use finite-element models in the design and development of products such as aircraft wings and turbo-machinery. Finite-element modeling has been applied in areas such as:

- thermal problems;
- electromagnetic problems;
- fluid problems; and
- structural modeling.

For example, in structural mechanics, finite-element models have produced stiffness and strength visualizations as well as minimized weight, materials and costs. Finite-element modeling usually consists of the following essential steps (Chandrupatla and Belegudu, 2002):

- choosing elements; and
- choosing the basis functions.

In finite-element analysis, a computer model is developed to analyze a structure and this model is used in areas such as new product design or on improving the performance of existing products. This allows engineers to know in advance if a design will perform to the required specifications before the manufacturing process is commenced. Normally, there are two kinds of finite-element analysis that are used. These are (Solin *et al.*, 2004):

- 2-dimensional modeling; and
- 3-dimensional modeling.

Even though 2-dimensional modeling is simple and permits computationally efficient analysis, it gives reduced accuracy. Results that are more accurate can be obtained through 3-dimensional modeling. However, this is computationally expensive. Furthermore, finite-element analysis can be formulated such that the system is linear or nonlinear. Modeling a linear system is not as complex and usually does not consider plastic deformation, while nonlinear systems do take plastic deformation into account. In this book, we deal with linear finite-element modeling represented by a second-order ordinary differential equation that consists of mass, damping and stiffness matrices.

Finite-element analysis uses a system of points called *nodes*, which form a grid known as a *mesh* as shown in Figure 1.1. The mesh is modeled to include the material and structural properties that describe the manner in which the structure will respond to particular loading and boundary conditions.

In Figure 1.1, a finite-element model was constructed using ABAQUS (1994) to study the dynamics of the cylinders. The cylinder has a diameter of 100 mm, a height of 100 mm and a thickness of 1.75 mm. This finite-element model consists of 1001 8-noded-shell-elements and 4100 nodes. This size of elements was chosen because it was found that increasing the mesh size did not improve the results any further, implying that the finite-element model had converged. This figure shows the mode shape of the first natural frequency occurring at 433 Hz.

The process of adding a mass of 5 g at various positions in the finite-element model is followed to study the dynamics of the cylinder. It was observed that adding this mass to a cylinder that was symmetrical, breaks down the symmetry, thereby eliminating the incidence of repeated modes. (The initial mass of the cylinder was 0.43 kg.)

These loaded nodes are allocated with a particular density throughout the material, according to the expected stress levels of that area (Baran, 1988). Sections that experience a great deal of stress will then normally have a higher node density than those that encounter slight or no stress. Points of stress concentration may contain fracture points of formerly tested material, joints, welds and high stress areas. The mesh may be visualized as a spider web so that from each node, a mesh element broadens to each of the neighboring nodes. This web of vectors carries the material properties of the object, therefore making many elements to be studied.

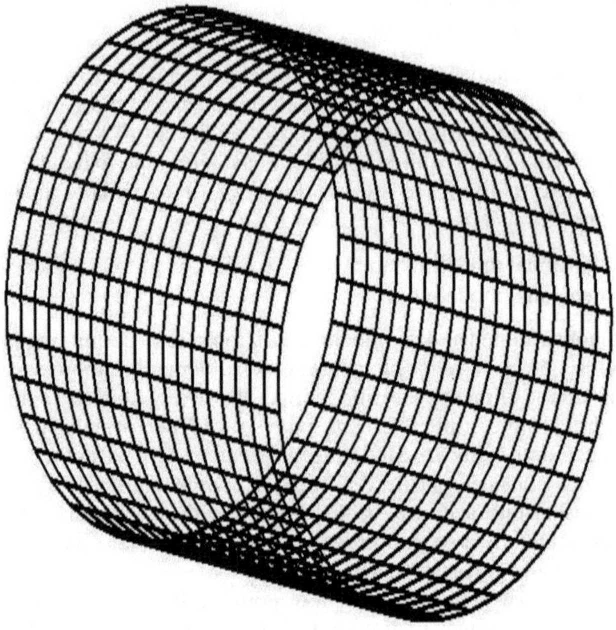

Figure 1.1 A finite-element model of a cylindrical shell (Marwala, 2001)

On implementing finite-element modeling, a choice of elements needs to be made and these include: beam, plate, shell and solid elements.

Pertinent questions that need to be answered when implementing finite-element models include: is the material isotropic (identical throughout material), orthotropic (only identical at 90°) or anisotropic (different throughout the material) (Irons and Shrive, 1983; Zienkiewicz, 1986)?

Finite-element analysis can be used to model a class of the following problems (Zienkiewicz, 1986):

- Vibration analysis, which is used to test a structure for random vibrations, impact and shock. In this analysis, issues such as natural frequencies and mode shapes are dealt with.
- Fatigue analysis, which aids the engineer to approximate the life-cycle of a material or a structure due to cyclical loading. This analysis can reveal the sections of the structure with a high probability of crack propagation.
- Heat-transfer analysis, which models the conductivity or thermal fluid dynamics of the material or structure.

Miao *et al.* (2009) successfully applied a 3-dimensional finite-element model for the simulation of shot peening, which is a cold-working process that is used primarily to extend the fatigue life of metallic components. Hlilou *et al.* (2009) used finite-element modeling for softening material behavior, while Pepper and Wang (2007) applied a finite-element model for wind-energy assessment of renewable energy in Nevada, White *et al.* (2008) applied a 3-dimensional unstructured mesh finite-element model for shallow-water modeling, while Zhang and Teo (2008) applied a finite-element model for the treatment of a lumbar degenerative disc disease. Other successful applications of finite-element modeling include metal powder compaction process (Rahman *et al.*, 2009), rock mechanics (Chen *et al.*, 2009), ferroelectric materials (Schrade *et al.*, 2007), and orthopedics (Easley *et al.*, 2007).

Now that this chapter has described finite-element modeling, the next stage is to see how to validate these models using experimental data. In this book, the analyses pursued further use vibration data, the subject of the next section.

1.3 Vibration Analysis

There are four major ways in which vibration data may be represented in the time, modal, frequency and time–frequency domains (Marwala, 2001). The process of measuring data is illustrated in Figure 1.2, while Figures 1.3, 1.4 and 1.5 show data in the time and frequency domain for the mode shape shown in Figure 1.1.

Figure 1.2 shows three major components of the measurement procedure employed:

- The excitation of the structure: a modal hammer is used to excite the structure *e.g.*, a cylinder or an electromagnetic shaker can be used to excite the cylinder.
- The sensing of the response: an accelerometer is used to measure the acceleration response.
- Data acquisition and processing: the data is amplified, filtered, converted from analog to digital format (*i.e.*, A/D converter) and, finally, stored in the computer.

In this book, we use data in the frequency domain. Raw data are measured in the time domain, and Fourier-transform techniques are used to transform data into the frequency domain. The modal properties are extracted from the frequency-domain data (and at times directly from the time domain). Theoretically, all of these domains include similar information, but in reality this is not automatically the situation. Since the time-domain data are complicated to understand, they are not used widely for fault identification. For this reason, this chapter reviews merely the modal and frequency domains.

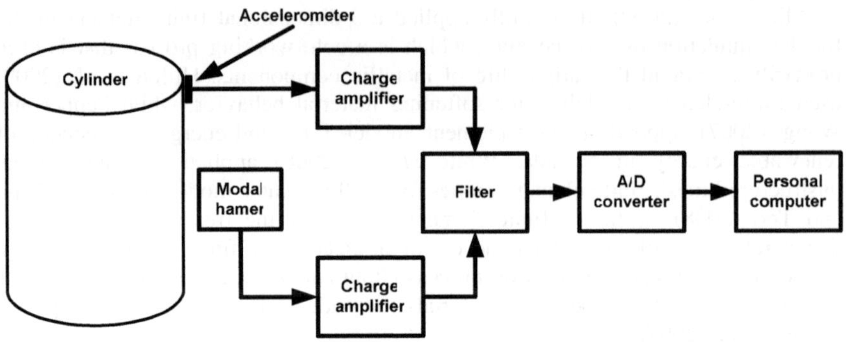

Figure 1.2 Schematic representation of a typical test set up

1.4 Domains Used for Finite-element-model Updating

1.4.1 Modal-domain Data (MDD)

The modal-domain data are expressed as natural frequencies, damping ratios and mode shapes. This book concentrates on natural frequencies and mode shapes because the systems in question are lightly damped. The most widely used technique for extracting the modal properties is the process called *modal analysis* (Ewins, 1995). The modal data have been used independently and in tandem for fault identification.

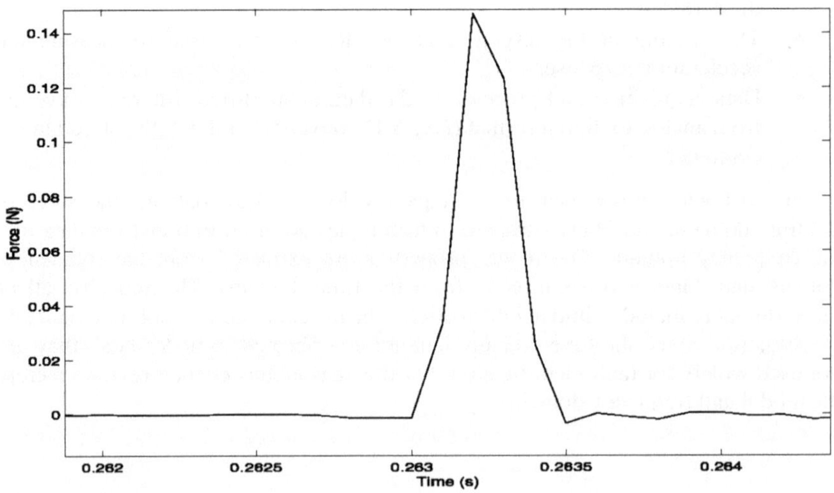

Figure 1.3 Impulse in time domain

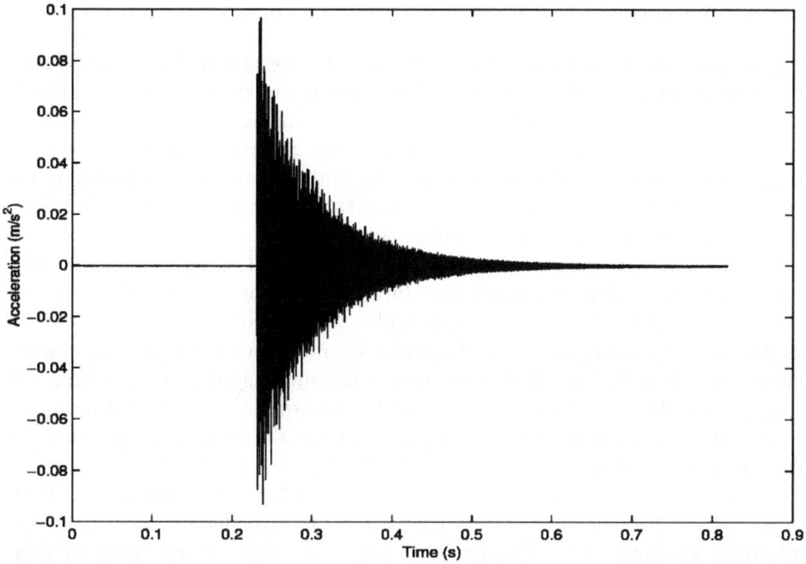

Figure 1.4 Response in time domain

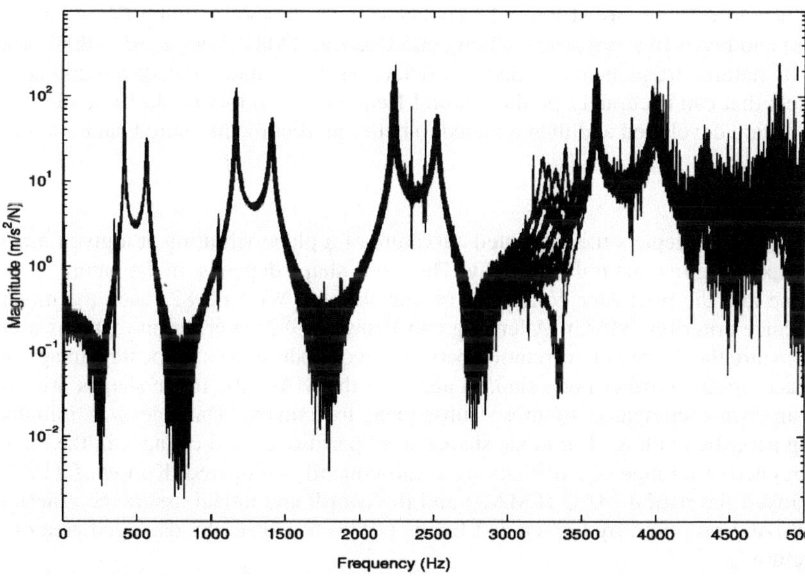

Figure 1.5 Frequency-response function which was obtained by dividing the Fourier transform of the responses by that of the impulse excitation

A) Natural Frequencies

Natural frequencies are fundamental properties of a system and can be revealed using vibration analysis. Shifts in natural frequencies have been used to identify structural damage. Cawley and Adams (1979) used changes in natural frequencies to identify damage in composite materials. To compute the ratio between frequency shifts for two modes, they regarded a grid between likely damage points and created an error term that related measured frequency shifts to those predicted by a model based on a local stiffness reduction.

Farrar *et al.* (1994) implemented the shifts in natural frequencies to identify damage on an I-40 bridge and noted that shifts in the natural frequencies were not adequate for detecting damage of small faults. To improve the accuracy of the natural-frequency technique, it was found to be more practical to carry out the experiment in controlled environments where the uncertainties of measurements were comparatively low. One example of such a controlled environment used is in using resonance ultrasound spectroscopy to measure the natural frequencies and establish the out-of-roundness of ball bearings (Migliori *et al.*, 1983).

Other successful usages of natural frequencies include (Messina *et al.*, 1996; Messina *et al.*, 1998) who successfully used the natural frequencies to locate single and multiple damages in a simulated 31-bar truss and tabular steel offshore platform. Damage was introduced to the two structures by reducing the stiffness of the individual bars by up to 30%. This technique was experimentally validated on an aluminum test-rod structure, where damage was introduced by reducing the cross-sectional area of one of the members from 7.9 to 5.0 mm.

Further applications of natural frequencies include spot welding (Wang *et al.*, 2008) and beam-like structures (Zhong and Oyadiji, 2008; Zhong *et al.*, 2008). The use of natural frequencies in damage detection necessitates the development of models that can accurately predict natural frequencies. In this book, finite-element models are developed and then updated to better predict the measured data.

B) Mode Shapes

A mode shape depicts the estimated curvature of a plane vibrating at a given mode corresponding to a natural frequency. The mode shape depends on the nature of the surface and the boundary conditions of that surface. West (1982) used the modal assurance criterion (MAC) (Allemang and Brown, 1982), a criterion that was used to measure the degree of correlation between two mode shapes to locate damage on a Space Shuttle Orbiter body flap. In applying the MAC, the mode shapes prior to damage were compared to those subsequent to damage. Damage was initiated using acoustic loading. The mode shapes were partitioned and changes in the mode shapes across a range of partitions were subsequently compared. Kim *et al.* (1992) employed the partial MAC (PMAC) and the coordinate modal assurance criterion (COMAC) proposed by Lieven and Ewins (1988) to isolate the damaged area of a structure.

Salawu (1995) established a global damage integrity index, based on a weighted ratio of the natural frequencies of damaged to undamaged structures. The weights were used to specify the sensitivity of each mode to damage. Steenackers and

Guillaume (2006) applied finite-element-model updating that took into account the uncertainty in the modal parameters.

Further applications of mode shapes include composite laminated plates by Araújo dos Santos (2006) as well as Qiao *et al.* (2007), linear structures by Fang and Perera (2009), beam-type structures by Qiao and Cao (2008), and other structures by Sazonov and Klinkhachorn (2005). The main drawbacks of the modal properties as outlined by Ewins (1995) are that they are:

- computationally expensive to identify because they involve some optimization procedure to identify them;
- vulnerable to added noise due to modal analysis;
- not capable of factoring the out-of-frequency-bandwidth modes; and
- merely appropriate for lightly damped and linear structures.

However, the modal data have the following advantages as outlined by Ewins (1995):

- simple to employ for damage identification;
- most appropriate for detecting large faults;
- directly associated with the topology of the structure; and
- emphatic of the significant aspects of the dynamics of the structure.

1.4.2 Frequency-domain Data

The measured excitation and response of the structure are transformed into the frequency domain using Fourier transforms (Ewins, 1995). This is shown in Figure 1.3. The ratio of the response to excitation in the frequency domain at each frequency is called the *frequency-response-function* (FRF). The direct use of the frequency-response functions without extracting the modal data to identify faults has become a subject of research (Sestieri and D'Ambrogio, 1989; Faverjon and Sinou, 2009). D'Ambrogio and Zobel (1994) directly applied the frequency-response functions to identify the presence of faults in a truss structure. Imregun *et al.* (1995) observed that the direct use of the frequency-response functions to categorize faults in simulated structures yields certain advantages over the use of modal properties. Lyon (1995) and Schultz *et al.* (1996) have advocated the use of measured frequency-response functions for structural diagnostics. Other direct applications of the frequency-response functions include Shone *et al.* (2009), Ni *et al.* (2006), Liu *et al.* (2009) and White *et al.* (2009), as well as Todorovska and Trifunac (2008). Frequency-response functions are difficult to use in that (Maia and Silva, 1997):

- they contain more information than is needed for damage identification;
- there is also no method to choose the frequency bandwidth of interest; and
- they are generally noisy in the anti-resonance regions.

Yet, FRF methods have the following advantages (Maia and Silva, 1997):

- measured data include the effects of out-of-frequency-bandwidth modes;

- one measurement provides abundant data;
- modal analysis is not required and, therefore, modal identification errors are avoided; and
- frequency-response functions are applicable to structures with high damping and modal density.

These methods have revealed several promises but extensive research is still required on how frequency-response functions can best be employed for fault identification. In addition to the modal data, other data that can be used for finite-element model updating include strain data (Yao and Li, 2008), or a combination of static displacement and modal data (Zong and Xia, 2008). In this section, two different domains in which vibration data may be presented were reviewed.

1.5 Finite-element-model Updating Methods

In real life, it turns out that the prediction of the finite-element model is quite different from the measurements. As an example, for a finite-element model of Figure 1.1, the differences between the model the predictions and measured results are shown in Table 1.1.

In this book, we investigate some of the updating methods that have been proposed and applied in the past so that suitable methods may be chosen. In particular, we focus our attention on those methods that are either based on computational intelligence or are inspired by computational intelligence.

Table 1.1 The comparison between finite-element model and measurements

Mode number	Finite-element frequencies (Hz)	Measured frequencies (Hz)
1	0433.3	0413.7
2	0445.5	0425.3
3	0587.5	0561.0
4	0599.0	0576.6
5	1218.3	1165.0
6	1262.9	1196.8
7	1480.0	1480.1
8	1510.0	1483.4
9	2273.5	2229.3
10	2323.6	2346.2
11	2422.3	2520.1
12	2657.4	2612.1
13	2711.3	–
14	2778.4	–
15	3713.7	3330.2
16	3914.3	3585.8
17	4138.5	3990.6
18	4222.8	4309.5
19	4634.2	4814.2

Finite-element-model updating has been used to detect damage in structures (Friswell and Mottershead, 1995; Mottershead and Friswell, 1993; Maia and Silva, 1997). As explained before, there are two approaches used in finite-element-model updating: direct methods and iterative methods.

Direct methods, which use the modal properties, are computationally efficient to implement and reproduce the measured modal data exactly. They do not take into account the physical parameters that are updated. Consequently, even though the finite-element-model can predict measured quantities, the updated model is limited in the following ways (Maia and Silva, 1997):

- it may be deficient in the connectivity of nodes – connectivity of nodes is a phenomenon that appears physically in finite-element modeling due to the certainty that the structure is physically connected;
- the updated matrices are populated instead of banded – the reality that structural elements are simply connected to their neighbors makes sure that the mass and stiffness matrices are diagonally dominated with few couplings between elements that are far apart; and
- there is a potential of the loss of symmetry of the matrices.

Iterative procedures use changes in physical parameters to update the finite-element models and, thereby, generate models that are physically realistic. Esfandiari *et al.* (2009) used frequency-response functions and natural frequencies for model updating in structures. A least-squares technique with suitable normalization was used for solving the over-determined system with noisy data. The sensitivity approach and appropriate choice of measured frequency data gave better accuracy and convergence of the finite-element model updating process.

Wang *et al.* (2009) used the Zernike moment descriptor (ZMD) for recognizing mode shapes and finite-element-model updating. When this approach was applied to mode-shape recognition problem for a simple plate structure, it was observed that the ZMD has substantial benefits when compared to the conventional modal assurance criterion (MAC), especially in axisymmetric structures.

Yuan and Dai (2009) developed an efficient numerical technique for the finite-element-model updating of damped gyroscopic systems. This method integrated the measured modal data into the finite-element model to construct an updated finite-element model that resulted in damping and gyroscopic matrices that strongly reproduced experimental modal data.

Kozak *et al.* (2009) used a miscorrelation index for model updating. An index known as the miscorrelation index (MCI) was established to pinpoint the degrees of freedom transmitting errors in a finite-element model. The MCI was computed from the frequency-response functions and the dynamic stiffness matrix for every coordinate as a function of frequency. The MCI sensitivity method gave good results even when only a few degrees of freedom were measured.

Arora *et al.* (2009) implemented finite-element-model updating that used damping matrices. The finite-element-model updating method based on damping was proposed and examined with the objective that the damped finite-element-updated model was capable of accurately predicting the measured frequency-response functions. The results obtained demonstrated that the proposed method could accurately predict complex frequency-response functions.

Schlune *et al.* (2009) implemented finite-element-model updating to improve bridge evaluation. Their method was intended to remove erroneous modeling over-simplification through physical model improvements prior to parameters being approximated by nonlinear optimization. Additionally, multi-response objective functions advanced and permitted the hybridization of different kinds of measurements to attain a consistent procedure for parameter estimation.

Yang *et al.* (2009) investigated several objective functions for finite-element-model updating in structures. Bayraktar *et al.*, (2009) applied modal properties to change uncertain material properties and boundary conditions and update finite-element models of a bridge. Further, Li and Du (2009) used the most sensitive design variable for finite-element-model updating of stay cables and successfully identified a finite-element model that could predict natural frequencies that were nearer to the experimental ones.

Steenackers *et al.* (2007) successfully applied transmissibility data (the ratio between two responses) for finite-element-model updating. Other successful implementations of finite-element-model updating methods include applications in bridges (Huang and Zhu, 2008; Jaishi *et al.*, 2007), composite floors (Pavic *et al.*, 2007), helicopters (Shahverdi *et al.*, 2006), atomic force microscopes (Chen, 2006) and in steel box-girder footbridges (Živanović *et al.*, 2007).

One important issue in finite-element-model updating is the issue of parameter selection. Kim and Park (2008) developed an automated parameter-selection method for finite-element-model updating. This automated parameter-selection method was based on straightforward observations. The effectiveness of the proposed method was positively evaluated on a simulated problem, as well as on a real engineering structure. Zárate and Caicedo (2008) studied the issue of multiple existences of updated finite-element models and observed that the global optimum solution to the difference between measured data and finite-element data is not necessarily the desired updated finite-element model. Another issue of importance is that there is always a mismatch between the measured mode shapes and the finite-element model predicted mode shapes in terms of the measured degrees of freedom. One of the reasons for this may be the difficulty in measuring rotational degrees of freedom.

Li *et al.* (2008) employed the Guyan reduction method to reduce the degrees-of-freedom of the finite-element model. The other issue is the ill-conditioning that takes place during the finite-element-updating process. Several methods have been proposed to deal with this issue, including the Bayesian approach (Marwala and Sibisi, 2005) and regularization (Friswell and Mottershead, 1995). Wu and Dai (2008) used the regularized Lanczos method for model updating.

In this section, direct and indirect techniques that use the frequency-response functions or modal properties for finite-element-model updating are presented.

A) Matrix-update Methods

Matrix-update techniques are based on the modification of structural model matrices, for example the mass, stiffness and damping matrices, to identify damage in structures (Baruch, 1978). They are implemented by minimizing the distance between analytical and measured matrices as follows (Friswell and Mottershead, 1995):

$$\{E\}_i = (-\omega_i^2[M] + j\omega_i[C] + [K])\{\phi\}_i \tag{1.1}$$

Here, M is a subscript for a measured quantity; $[M]$ is the mass matrix; $[C]$ is the damping matrix; $[K]$ is the stiffness matrix of the structure; $\{E\}_i$ is the error vector also called the residual force; $j=\sqrt{-1}$; ω_i is the ith natural frequency; and $\{\phi\}_i$ is the ith mode shape. In Equation 1.1 the residual force is the harmonic force with which the undamaged structure will have to be excited at a frequency of ω_i so that the structure will respond with the mode shape $\{\phi\}_i$. The Euclidean norm of $\{E\}_i$ is minimized by updating physical parameters in the model.

The difference between updated matrices and original matrices identifies the damage. One approach for implementing this procedure is to formulate the objective function to be minimized, place constraints on the problem such as retaining the orthogonal relations of the matrices (Ewins, 1995) and choosing an optimization routine. These techniques are classified as *iterative* since they are employed by iteratively changing the relevant parameters until the error is minimized.

Ojalvo and Pilon (1988) minimized the Euclidean norm of the residual force for the ith mode of the structure by using the modal properties. Yuan and Dai (2006) used measured incomplete modal data, maintaining the required orthogonal relations and the Frobenius approach for updating finite-elements. D'Ambrogio and Zobel (1994) minimized the residual force in the equation of motion in the frequency domain as (Friswell and Mottershead, 1995):

$$[E] = (-\omega^2[M] + j\omega[C] + [K])[X^m] - [F^m] \tag{1.2}$$

In Equation 1.2 $[X^m]$ and $[F^m]$ are the Fourier-transformed displacement and force matrices, respectively. Each column of the matrix corresponds to a measured frequency point. The Euclidean norm of the error matrix $[E]$ is minimized by updating physical parameters in the model. The methods described in this subsection are computationally expensive. In addition, it is difficult to find a global minimum through the optimization technique, due to the multiple stationary points, which are caused by its nonunique nature (Janter and Sas, 1990). Techniques such as the use of genetic algorithms and multiple starting design variables have been applied to increase the probability of finding the global minimum (Mares and Surace 1996; Levin and Lieven, 1998; Larson and Zimmerman, 1993; Friswell *et al.*, 1994; Dunn, 1998).

B) Optimal Matrix Methods

Optimal matrix techniques are classified as direct methods and employ analytical, rather than numerical solutions to obtain matrices from the damaged systems. They are normally formulated in terms of Lagrange multipliers and perturbation matrices. The optimization problem is posed to minimize (Friswell and Mottershead, 1995):

$$\{E([\Delta M],[\Delta C],[\Delta K]) + \lambda R([\Delta M],[\Delta C],[\Delta K]) \tag{1.3}$$

Here, E is the objective function; λ is the Lagrange multiplier; R is the constraint of the equation; and Δ is the perturbation of system matrices.

In Equation 1.3, different combinations of perturbations are experimented with until the difference, between the finite-element model and the measured results, is minimized. Baruch and Bar Itzhack (1978), Berman and Nagy (1983) and Kabe (1985) formulated Equation 1.3 by minimizing the Frobenius norm of the error while retaining the symmetry of the matrices. McGowan *et al.* (1990) introduced an additional constraint that maintained the connectivity of the structure and used measured mode shapes to update the stiffness matrix to locate structural damage. Zimmerman *et al.* (1995) used a partitioning technique for matrix perturbations as sums of element or substructural perturbation matrices to reduce the rank of unknown perturbation matrices. The result was a reduction in the modes required to successfully locate damage. Carvalho *et al.* (2007) successfully applied a direct technique for model updating with incomplete measured modal data. One limitation of these methods is that the updated model is not always physically realistic.

C) Sensitivity-Based Methods

Sensitivity-based methods assume that experimental data are perturbations of design data about the original finite-element model. Owing to this assumption, experimental data must be close to the finite-element data for these methods to be effective. This formulation only works if the structural modification is small, that is, the magnitude of damage is small.

These methods are based on the calculation of the derivatives of either the modal properties or the frequency-response functions. There are many methods that have been developed to calculate the derivative of the modal properties and frequency-response functions. One such method was proposed by Fox and Kapoor (1968) who calculated the derivatives of the modal properties of an undamped system. Norris and Meirovitch (1989), Haug and Choi (1984), Chen and Garba (1980) put forward other methods of computing the derivatives of the modal properties to ascertain parameter changes. They used orthogonal relations with respect to the mass and stiffness matrices to compute the derivatives of the natural frequencies and mode shapes with respect to parameter changes.

Ben-Haim and Prells (1993) proposed selective frequency-response function sensitivity to uncouple the finite-element-updating problem, while Lin *et al.* (1995) improved the modal sensitivity technique by ensuring that these methods were applicable to large magnitude faults.

Hemez (1993) proposed a method that assesses the sensitivity at an element level. The advantage of this method is its ability to identify local errors. In addition, it is computationally efficient. Alvin (1996) modified this approach to improve the convergence rate by using a more realistic error indicator and by incorporating statistical confidence measurements for initial model parameters and measured data.

D) Eigenstructure-assignment Methods

Eigenstructure-assignment methods are based on control-system theory. The structure under investigation is forced to respond in a predetermined manner. During damage detection, the desired eigenstructure is the one that is measured in the test. Zimmerman and Kaouk (1992) applied these methods to identify the elastic modulus of a cantilevered beam using measured modal data. Schultz *et al.* (1996) improved this approach through using measured frequency-response functions.

The one limitation of the methods outlined in this section is that the number of sensor locations is less than the degrees of freedom in the finite-element model. This is problematic since it makes difficult the integration of the experimental data and finite-element model, the very basis of finite-element updating fault-identification methods. To compensate for this limitation, the mode shapes and frequency-response functions are either expanded to the size of the finite-element model or the mass and stiffness matrices of the finite-element model are reduced to the size of the measured data.

The reduction methods that have been used include static reduction (Guyan, 1965), dynamic reduction (Paz, 1984), improved reduced system (O'Callahan, 1989) and the system-equivalent-reduction process (O'Callahan *et al.*, 1989). The system-equivalent-expansion process was used to expand the measured mode shapes and frequency-response functions. Techniques that expand the mass and stiffness matrices have also been employed (Gysin, 1990; Imregun and Ewins, 1993; Friswell and Mottershead, 1995).

It has been shown that finite-element-updating techniques have numerous limitations. Most importantly, they rely on an accurate finite-element model, which may not be available. Even if the model is available, the problem of the nonuniqueness of the updated model makes the problem of damage identification using finite-element updating nonunique. Nonuniqueness is a phenomenon that describes a situation where more than one updated finite-element model is used.

E) Iterative Optimization Methods

Huang and Zhu (2008) applied optimization methods for the finite-element-model updating of bridge structures. The optimization method was augmented by a sensitivity analysis. Schwarz *et al.* (2007) updated a finite-element model that minimized the difference between the modes of a finite-element model and those from the experiment. Bakir *et al.* (2007) applied sensitivity approaches for finite-element-model updating. They used a constrained optimization method to minimize the differences between the natural frequencies and mode shape.

Jaishi and Ren (2007) applied a multi-objective optimization approach for finite-element-model updating. Their multi-objective cost function was based on the differences between eigenvalues and strain energy. Liu *et al.* (2006) updated a finite-element model of a 14-bay beam with semirigid joints and a boundary using a hybrid optimization method. Zhang and Huang (2008) applied a gradient-descent optimization method for the finite-element-model updating of bridge structures. The objective function was formulated as the summation of the frequency difference and modal shapes. Parameter alteration was guided by engineering judgment.

F) Bayesian / Monte Carlo Approaches

A Bayesian approach is a procedure based on Bayes' Theorem and functions for conducting statistical inference through using the evidence (observations) to update the probability that a hypothesis may be true (Marwala, 2009). Wong *et al.* (2006) applied Bayesian approaches for updating a bridge model using sensor data, while Marwala and Sibisi (2005) applied finite-element updating in beam structures. Mares *et al.* (2006) successfully applied Monte Carlo method for stochastic model updating. Zheng *et al.* (2009) applied a Bayesian approach for the finite-element-model updating of a long-span, steel sky-bridge. Hemez and Doebling (1999) successfully applied a Bayesian approach to finite-element-model updating and applied this to linear dynamics, while Lindholm and West (1995) applied a Bayesian parameter approximation for finite-element-model updating and applied this to model experimental dynamic response data.

G) Computational Intelligence Methods

Liu *et al.* (2009) applied fuzzy theory for finite-element-model updating. In their research, the model parameters and design variables were modeled as fuzzy variables and this technique was successfully implemented on an actual concrete bridge. Jung and Kim (2009) implemented a hybrid genetic algorithm for finite-element-model updating and tested this procedure on a numerical bridge model. A hybrid genetic algorithm was formed by combining a genetic algorithm with Nelder–Mead simplex method. The proposed technique was found to be effective on the finite-element updating of bridge structures.

Tan *et al.* (2009) applied support vector machines and wavelet data for finite-element-model updating in structures. The result obtained from simulated data validated that this approach could successfully update the model. Zapico *et al.* (2008) applied neural networks for finite-element updating. The results obtained showed that the updated finite-element model could accurately predict the low modes that were identified from measurements. Other successful applications of computational intelligence techniques to finite-element updating include Tu and Lu (2008), Yan *et al.* (2007) as well as Fei *et al.* (2006) who applied genetic algorithms, Feng *et al.* (2006) who applied a hybrid of genetic algorithm and simulated annealing, and He *et al.* (2008) who applied a hybrid of genetic algorithm and neural networks.

1.6 Computational Intelligence Methods

From earlier work on finite-element-model updating it is evident that finite-element-model updating is essentially an optimization method. Here, the design variables are the uncertain parameters in the model. The objective is to minimize the distance between the finite-element predicted data and the measured data. Some of the optimization methods implemented in this study are outlined below.

The Nelder–Mead simplex method, a nongradient-based technique, intended to solve standard unconstrained optimization problem of minimizing a particular nonlinear function. It uses function values at several positions and does not attempt to calculate an estimated gradient at any of these positions (Bürmen *et al.*, 2006).

The sequential quadratic programming technique, which solves a sequence of subproblems aimed at minimizing a quadratic representation of the objective function. When the problem is unconstrained, then the technique becomes Newton's technique (Boggs and Tolle, 1995).

A genetic algorithm is a simulation of natural evolution where the law of the survival of the fittest is applied to a population of individuals. This natural optimization method is used for optimizing a function (Mitchell, 1998).

Particle-swarm optimization is an evolutionary optimization method that was developed by Kennedy and Eberhart (1995) and was inspired by algorithms that model the "flocking behavior" seen in birds.

Simulated annealing is a Monte Carlo method that is used to investigate the equations of state and frozen states of n degree-of-freedom system and can be used to solve an optimization problem (Kirkpatrick *et al.*, 1983).

The response-surface method is a procedure that functions by generating a response for a given input and then constructs an approximation to a complicated model such as a finite-element model (Kamrani *et al.*, 2009).

Hybrid optimization methods are types of optimization methods that combine more than one algorithm (Li and Yu, 2009).

The multi-objective optimization method is an optimization method that makes use of more than one objective function (Soares and Vieira, 2009).

A Bayesian approach is a parameter-estimation technique that can be trained using a Monte Carlo method (Yuan *et al.*, 2009).

Finite-element models are computationally expensive methods. To manage the computational load, some form of emulator to approximate the finite-element model can be implemented. The emulators implemented in this book are:

- a multi-layer perceptron neural network, which is a feedforward network, consisting of linear or nonlinear activation functions, that is used to map sets of inputs to outputs (Sancho-Gómez *et al.*, 2009); and
- a radial basis function network, which is a type of feedforward network that has radial basis functions as its activation functions (Yeung *et al.*, 2009).

1.7 Outline of the Book

In Chapter 2, the Nelder–Mead simplex and Broyden–Fletcher–Goldfarb–Shanno (BFGS) optimization methods are introduced, applied and compared for finite-element model updating. The use of reduction and expansion methods to equate measured modal data to finite-element systems matrices is also investigated.

In Chapter 3, genetic algorithms are introduced and applied to finite-element-model updating. This approach is compared to the Nelder–Mead simplex method. In Chapter 4, the particle-swarm optimization method is introduced and applied to finite-element-model updating and the results are compared to those obtained from the genetic algorithm. In Chapter 5, simulated annealing is introduced and applied to finite-element-model updating and the results are compared to those from particle-swarm optimization. In Chapter 6, a response-surface method that combines the multi-layer perceptron and particle-swarm optimization is introduced and applied to finite-element-model updating. The results are compared to those from a genetic algorithm, particle-swarm optimization and simulated annealing. In Chapter 7, a hybrid optimization method is introduced that combines particle-swarm optimization, with the Nelder–Mead simplex method and it is applied to finite-element-model updating. The results are compared to those from when a genetic algorithm, particle-swarm optimization and simulated annealing are used individually. In Chapter 8, a multi-objective optimization method that uses both modal properties data and frequency-domain data is introduced for finite-element-model updating. Chapter 9 implements the multi-layer perceptron for finite-element-model updating, while Chapter 10 implements the Bayesian approach to finite-element-model updating. In Chapter 11, finite-element-model updating is applied for damage detection in structures. Finally, Chapter 12 concludes the book with key recommendations and outstanding issues for further development.

References

ABAQUS: (1994) ABAQUS Users Manual (Ver 5.4). Hibbitt, Karlsson and Sorensen, Inc, Pawtucket, Rhode Island

Allemang RJ, Brown DL (1982) A Correlation Coefficient for Modal Vector Analysis. In: Proc of the 1st Intl Modal Anal Conf:01–18

Alvin KF (1996) Finite Element Model Updating via Bayesian Estimation and Minimisation of Dynamic Residuals. In: Proc of the 14th Intl Modal Anal Conf:428–431

Araújo dos Santos JV, Lopes HMR, Vaz M, Mota Soares CM, Mota Soares CA, de Freitas MJM (2006) Damage Localization in Laminated Composite Plates Using Mode Shapes Measured by Pulsed TV Holography. Compos Struct 76:272–281

Arora V, Singh SP, Kundra TK (2009) Finite Element Model Updating with Damping Identification (2009) J of Sound and Vib 324:1111–1123

Babuska I, Banerjee U, Osborn JE (2004) Generalized Finite Element Methods: Main Ideas, Results, and Perspective. Intl J of Comput Methods 1:67–103

Bakir PG, Reynders E, De Roeck G (2007) Sensitivity-based Finite Element Model Updating Using Constrained Optimization with a Trust Region Algorithm. J of Sound and Vib 305:211–225

Baran NM (1988) Finite Element Analysis on Microcomputers. New York: McGraw-Hill Book Company

Baruch M (1978) Optimisation Procedure to Correct Stiffness and Flexibility Matrices Using Vibration Data. Am Inst of Aeronaut and Astronaut J 16:1208–1210

Baruch M, Bar Itzhack IY (1978) Optimum Weighted Orthogonalization of Measured Modes. Am Inst of Aeronaut and Astronaut J 20:1623–1626

Bayraktar A, Altunişik AC, Sevim B, Türker T (2009) Finite Element Model Updating of Kömürhan Highway Bridge [Kömürhan Köprüsünün Sonlu Eleman Model Iyileştirilmesi]. Teknik Dergi / Tech J of Turk Chamb of Civ Eng 20:4675–4700

Ben-Haim Y, Prells U (1993) Selective Sensitivity in the Frequency Domain, Part I: Theory. Mech Syst and Signal Process 7:461–475

Berman A, Nagy EJ (1983) Improvement of Large Analytical Model Using Test Data. Am Inst of Aeronaut and Astronaut J 21:1168–1173

Boggs PT, Tolle JW (1995) Sequential Quadratic Programming. Acta Numerica 4:01–51

Bürmen A, Puhan J, Tuma T (2006) Grid Restrained Nelder-Mead Algorithm. Comput Optim Appl 34:359–375

Carvalho J, Datta BN, Gupta A, Lagadapati M (2007) A Direct Method for Model Updating with Incomplete Measured Data and Without Spurious Modes. Mech Syst and Signal Process 21:2715–2731

Cawley P, Adams RD (1979) The Location of Defects from Measurements of Natural Frequencies. J of Strain Anal 14:49–57

Chandrupatla TR, Belegudu AD (2002) Introduction to Finite Elements in Engineering. Prentice Hall, New Jersey

Chen JC, Garba JA (1980) Analytical Model Improvement Using Modal Test Results. Am Inst of Aeronaut and Astronaut J 18:684–690

Chen KN (2006) Model Updating and Optimum Designs for V-shaped Atomic Force Microscope Probes. Eng Optim 38:755–770

Chen S, Fu C, Isam S (2009) Finite Element Analysis of Jointed Rock Masses Reinforced by Fully-grouted Bolts and Shotcrete Lining. Intl J of Rock Mech and Min Sci 46:19–30

Courant R, Robbins H (1941) What is Mathematics? Oxford University Press, New York

D'Ambrogio W, Zobel PB (1994) Damage Detection in Truss Structures Using a Direct Updating Technique. In: Proc of the 19th Intl Seminar for Modal Anal:657–667

Dunn SA (1998) The Use of Genetic Algorithms and Stochastic Hill-climbing in Dynamic Finite-element Model Identification. Comput and Struct 66:489–497

Easley SK, Pal S, Tomaszewski PR, Petrella AJ, Rullkoetter PJ, Laz PJ (2007) Finite Element-based Probabilistic Analysis Tool for Orthopaedic Applications. Comput Methods and Programs in Biomed 85:32–40

Esfandiari A, Bakhtiari-Nejad F, Rahai A, Sanayei M (2009) Structural Model Updating Using Frequency Response Function and Quasi-linear Sensitivity Equation (2009) J of Sound and Vib 326:557–573

Ewins DJ (1995) Modal Testing: Theory and Practice. Research Studies Press, Letchworth

Fang S, Perera R (2009) Power Mode Shapes for Early Damage Detection in Linear Structures. J of Sound and Vib 324:40–56

Farrar CR, Baker WE, Bell TM, Cone KM, Darling TW, Duffey TA, Eklund A, Migliori A (1994) Dynamic Characteristics and Damage Detection in the I-40 Bridge over the Rio Grande. Los Alamos National Laboratory Report LA-12767-MS

Faverjon B, Sinou JJ (2009) Identification of an Open Crack in a Beam Using a Posteriori Error Estimator of the Frequency Response Functions with Noisy Measurements. Eur J of Mech – A / Solids 28:75–85

Fei Q, Li A, Miao C (2006) Dynamic Finite Element Model Updating Using Meta-model and Genetic Algorithm. J of Southeast Univ (Engl Ed) 22:213–217

Feng FZ, Kim YH, Yang BS (2006) Applications of Hybrid Optimization Techniques for Model Updating of Rotor Shafts. Struct and Multidiscip Optim 32:65–75

Fox RL, Kapoor MP (1968) Rates of Change of Eigenvalues and Eigenvectors. Am Inst of Aeronaut and Astronaut J 6:2426–2429

Friswell MI, Mottershead JE (1995) Finite Element Model Updating in Structural Dynamics. Kluwer Academic Publishers Group, Norwell

Friswell MI, Penny JET, Wilson DAL (1994) Using Vibration Data and Statistical Measures to Locate Damage in Structures. Modal Anal: The Intl J of Anal and Exp Modal Anal 9:239–254

Guyan RJ (1965) Reduction of Stiffness and Mass Matrices. Am Inst of Aeronaut and Astronaut J 3:380

Gysin H (1990) Comparison of Expansion Methods for FE Model Localization. In: Proc of the 8th Intl Modal Anal Conf:195–204

Hastings JK, Juds MA, Brauer JR (1985) Accuracy and Economy of Finite Element Magnetic Analysis. 33rd Annu Natl Relay Conf:45–50

Haug EF, Choi KK (1984) Structural Design Sensitivity with Generalized Global Stiffness and Mass Matrices. Am Inst of Aeronaut and Astronaut J 22:1299–1303

He HX, Yan WM, Wang Z (2008) Stepwise Model Updating Method Based on Substructures and GA-ANN. Eng Mech 25:99–105

Hemez FM (1993) Theoretical and Experimental Correlation between Finite Element Models and Modal Tests in the Context of Large Flexible Structures. PhD Dissertation, University of Colorado, Boulder

Hemez FM, Doebling SW (1999) Validation of Bayesian Finite Element Model Updating for Linear Dynamics. In: Proc of the Intl Modal Anal Conf - IMAC 2: 1545–1555

Hlilou A, Ben Naceur I, Saï K, Gérard C, Forest S, Cailletaud G (2009) Generalization of the Polycrystalline β-model: Finite Element Assessment and Application to Softening Material Behavior. Comput Mater Sci 45:1104–1112

Hrennikoff A (1941) Solution of Problems of Elasticity by the Frame-Work Method. ASME J Appl Mech 8:A619–A715

Huang M, Zhu H (2008) Finite Element Model Updating of Bridge Structures Based on Sensitivity Analysis and Optimization Algorithm. Wuhan Univ J of Nat Sci 13:87–92

Imregun M, Ewins DJ (1993) An Investigation into Mode Shape Expansion Techniques. In: Proc of the 11th Intl Modal Anal Conf:168–175

Imregun M, Visser WJ, Ewins DJ (1995) Finite Element Model Updating Using Frequency Response Function Data-I. Theory and Initial Investigation. Mech Syst and Signal Process 9:187–202

Irons B, Shrive N (1983) Finite Element Primer. John Wiley

Jaishi B, Kim HJ, Kim MK, Ren WX, Lee SH (2007) Finite Element Model Updating of Concrete-filled Steel Tubular Arch Bridge Under Operational Condition Using Modal Flexibility. Mech Syst and Signal Process 21:2406–2426

Jaishi B, Ren WX (2007) Finite Element Model Updating Based on Eigenvalue and Strain Energy Residuals Using Multiobjective Optimization Technique. Mech Syst and Signal Process 21:2295–2317

Janter T, Sas P (1990) Uniqueness Aspects of Model-updating Procedure. Am Inst of Aeronaut and Astronaut J 28:538–543

Jung DS, Kim CY (2009) FE Model Updating Based on Hybrid Genetic Algorithm and its Verification on Numerical Bridge Model. Struct Eng and Mech 32:667–683

Kabe AM (1985) Stiffness Matrix Adjustment Using Mode Data. Am Inst of Aeronaut and Astronaut J 23:1431–1436

Kamrani B, Berbyuk V, Wäppling D, Stickelmann U, Feng X (2009) Optimal Robot Placement Using Response Surface Method. Intl J of Adv Manufact Technol 44:201–210

Kennedy JE, Eberhart RC (1995) Particle Swarm Optimization. In: Proc of the IEEE Intl Conf on Neural Netw:942–1948

Kim GH, Park YS (2008) An Automated Parameter Selection Procedure for Finite-element Model Updating and its Applications. J of Sound and Vib 309:778–793

Kim JH, Jeon HS, Lee SW (1992) Application of Modal Assurance Criteria for Detecting and Locating Structural Faults. In: Proc of the 10th Intl Modal Anal Conf:536–540

Kirkpatrick S, Gelatt CD, Vecchi MP (1983) Optimization by Simulated Annealing. Sci 220:671–680

Kozak MT, Öztürk M, Özgüven HN (2009) A Method in Model Updating Using Miscorrelation Index sensitivity. Mech Syst and Signal Process 23:1747–1758

Larson CB, Zimmerman DC (1993) Structural Model Refinement Using a Genetic Algorithm Approach. In: Proc of the 11th Intl Modal Anal Conf:1095–1101

Levin RI, Lieven NAJ (1998) Dynamic Finite Element Updating Using Neural Networks. J of Sound and Vib 210:593–608

Li H, Liu F, Hu SLJ (2008) Employing Incomplete Complex Modes for Model Updating and Damage Detection of Damped Structures. Sci in Chin, Ser E: Technol Sci 51:2254–2268

Li YF, Yu HB (2009) Hybrid Search Method for Efficiency Optimization Control of Induction Motor Drives. Elect Mach and Control 13:337–341

Li YQ, Du YL (2009) Dynamic Finite Element Model Updating of Stay-cable Based on the Most Sensitive Design Variable. J of Vib and Shock 28:141–143

Lieven NAJ, Ewins DJ (1988) Spatial Correlation of Mode Shapes, the Co-ordinate Modal Assurance Criterion. In: Proc of the 6th Intl Modal Anal Conf:690–695

Lin RM, Lim MK, Du H (1995) Improved Inverse Eigensensitivity Method for Structural Analytical Model Updating. J of Vib and Acoust 117:192–198

Lindholm BE, West RL (1995) Updating Finite Element Models with Experimental Dynamic Response Data using Bayesian Parameter Estimation. 1995 Collect of Tech Pap - AIAA/ASME/ASCE/AHS/ASC Struct, Struct Dyn and Mater Conf 2:794–802

Liu X, Lieven NAJ, Escamilla-Ambrosio PJ (2009) Frequency Response Function Shape-based Methods for Structural Damage Localization. Mech Syst and Signal Process 23:1243–1259

Liu Y, Duan Z, Liu H (2006) Updating Finite Element Model of Structures with Semi-rigid Joints and Boundary. In: Proc of SPIE - The Intl Soc for Opt Eng 6174 II: Art No 61743L

Liu Y, Duan Z, Liu H (2009) Updating of Finite Element Model in Considering Mode Errors with Fuzzy Theory. Key Eng Mater, 413–414:785–792

Lyon R (1995) Structural Diagnostics Using Vibration Transfer Functions. Sound and Vib 29:28–31

Maia NMM, Silva JMM (1997) Theoretical and Experimental Modal Analysis. Research Studies Press, Letchworth.

Mares C, Mottershead JE, Friswell MI (2006) Stochastic Model Updating: Part 1-Theory and Simulated Example. Mech Syst and Signal Process 20:1674–1695

Mares C, Surace C (1996) An Application of Genetic Algorithms to Identify Damage in Elastic Structures. J of Sound and Vib 195:195–215

Marwala T (2001) Fault Identification Using Neural Networks and Vibration Data. University of Cambridge. Unpublished Doctoral Thesis

Marwala T (2009) Computational Intelligence for Missing Data Imputation, Estimation and Management: Knowledge Optimization Techniques. Information Science IGI Global Publications, New York: Reference Imprint

Marwala T, Sibisi S (2005) Finite Element Updating Using Bayesian Framework and Modal Properties. J of Aircr 42:275–278

McGowan PE, Smith SW, Javeed M (1990) Experiments for Locating Damage Members in a Truss Structure. In: Proc of the 2nd USAF/NASA Workshop on Syst Identif and Health Monit of Precis Space Struct:571–615

Messina A, Jones IA, Williams EJ (1996) Damage Detection and Localization Using Natural Frequency Changes. In: Proc of the 1st Intl Conf on Identif in Eng Syst:67–76

Messina A, Williams EJ, Contursi T (1998) Structural Damage Detection by a Sensitivity and Statistical-based Method. J of Sound and Vib 216:791–808

Miao HY, Larose S, Perron C, Lévesque M (2009) On the Potential Applications of a 3D Random Finite Element Model for the Simulation of Shot Peening. Adv in Eng Softw 40:1023–1038

Migliori A, Bell TM, Dixon RD, Strong R (1983) Resonant Ultrasound Nondestructive Inspection. Los Alamos National Laboratory Report LS-UR-93-225

Mitchell M (1998) An Introduction to Genetic Algorithms (Complex Adaptive Systems), The MIT Press, Cambridge

Mottershead JE, Friswell MI (1993) Model Updating in Structural Dynamics: A Survey. J of Sound and Vib 167:347–375

Ni YQ, Zhou XT, Ko JM (2006) Experimental Investigation of Seismic Damage Identification Using PCA-compressed Frequency Response Functions and Neural Networks. J of Sound and Vib 290:242–263

Norris MA, Meirovitch L (1989) On the Problem of Modelling for Parameter Identification in Distributed Structures. Intl J for Numer Methods in Eng 28: 2451–2463

O'Callahan JC (1989) A Procedure for Improved Reduced System (IRS) Model. In: Proc of the 7th Intl Modal Anal Conf:17–21

O'Callahan JC, Avitabile P, Riemer R (1989) System Equivalent Reduction Expansion Process. In: Proc of the 7th Intl Modal Anal Conf:17–21

Ojalvo IU, Pilon D (1988) Diagnosis for Geometrically Locating Structural Mathematical Model Errors from Modal Test Data. In: Proc of the 29th AIAA / ASME / ASCE / AHS / ASC Struct, Struct Dynam, and Mater Conf:1174–1186

Pavic A, Miskovic Z, Reynolds P (2007) Modal Testing and Finite-element Model Updating of a Lively Open-plan Composite Building Floor. J of Struct Eng 133:550–558

Paz M (1984) Dynamic Condensation. Am Inst of Aeronaut and Astronaut J 22:724–727

Pepper W, Wang X (2007) Application of an H-adaptive Finite Element Model for Wind Energy Assessment in Nevada Renewable Energy 32:1705–1722

Qiao P, Cao M (2008) Waveform Fractal Dimension for Mode Shape-based Damage Identification of Beam-type Structures. Intl J of Solids and Struct 45:5946–5961

Qiao P, Lu K, Lestari W, Wang J (2007) Curvature Mode Shape-based Damage Detection in Composite Laminated Plates. Compos Struct 80:409–428

Rahman MM, Ariffin AK, Nor SSM (2009) Development of a Finite Element Model of Metal Powder Compaction Process at Elevated Temperature. Appl Math Model 33:4031–4048

Salawu OS (1995) Non-destructive Assessment of Structures Using Integrity Index Method Applied to a Concrete Highway Bridge. Insight 37:875–878

Sancho-Gómez JL, García-Laencina PJ, Figueiras-Vidal AR (2009) Combining Missing Data Imputation and Pattern Classification in a Multi-layer Perceptron. Intelli Autom and Soft Comput 15:539–553

Sazonov E, Klinkhachorn P (2005) Optimal Spatial Sampling Interval for Damage Detection by Curvature or Strain Energy Mode Shapes. J of Sound and Vib 285:783–801

Schlune H, Plos M, Gylltoft K (2009) Improved Bridge Evaluation Through Finite Element Model Updating Using Static and Dynamic Measurements. Eng Struct 31:1477–1485

Schrade D, Mueller R, Xu BX, Gross D (2007) Domain Evolution in Ferroelectric Materials: A Continuum Phase Field Model and Finite Element Implementation. Comput Methods in Appl Mech and Eng 196:4365–4374

Schultz MJ, Pai PF, Abdelnaser AS (1996) Frequency Response Function Assignment Technique for Structural Damage Identification. In: Proc of the 14th Intl Modal Anal Conf:105–111

Schwarz B, Richardson M, Formenti DL (2007) FEA Model Updating Using SDM Sound and Vib 41:18–23

Sestieri A, D'Ambrogio W (1989) Why be Modal: How to Avoid the Use of Modes in the Modification of Vibrating Systems. In: Proc of the 7th Intl Modal Anal Conf:25–30

Shahverdi H, Mares C, Wang W, Greaves CH, Mottershead JE (2006) Finite Element Model Updating of Large Structures by the Clustering of Parameter Sensitivities. Appl Mech and Mater 5–6:85–92

Shone SP, Mace BR, Waters TP (2009) Locating Damage in Waveguides from the Phase of Point Frequency Response Measurements. Mech Syst and Signal Process 23:405–414

Soares MM, Vieira GE (2009) A New Multi-objective Optimization Method for Master Production Scheduling Problems Based on Genetic Algorithm. Intl J of Adv Manufact Technol 41:549–567

Solin P, Segeth K, Dolezel I (2004) Higher-Order Finite Element Methods. Boca Raton: Chapman & Hall / CRC Press

Steenackers G, Devriendt C, Guillaume P (2007) On the Use of Transmissibility Measurements for Finite Element Model Updating. J of Sound and Vib 303:707–722

Steenackers G, Guillaume P (2006) Finite Element Model Updating Taking into Account the Uncertainty on the Modal Parameters Estimates. J of Sound and Vib 296:919–934

Tan D, Qu W, Wang J (2009) The Finite Element Model Updating of Structure Based on Wavelet Packet Analysis and Support Vector Machines. J of Huazhong Univ of Sci and Technol (Nat Sci Ed) 37:104–107

Todorovska MI, Trifunac MD (2008) Earthquake Damage Detection in the Imperial County Services Building III: Analysis of Wave Travel Times via Impulse Response Functions. Soil Dyn and Earthq Eng 28:387–404

Tu Z, Lu Y (2008) FE Model Updating Using Artificial Boundary Conditions with Genetic Algorithms. Comput and Struct 86:714–727

Wang R, Shang D, Li L, Li C (2008) Fatigue Damage Model Based on the Natural Frequency Changes for Spot-welded Joints. Intl J of Fatigue 30:1047–1055

Wang W, Mottershead JE, Mares C (2009) Mode-shape Recognition and Finite Element Model Updating Using the Zernike Moment Descriptor. Mech Syst and Signal Process 23:2088–2112

West WM (1982) Single Point Random Modal Test Technology Application to Failure Detection. The Shock and Vib Bull 52:25–31

White C, Li HCH, Whittingham B, Herszberg I, Mouritz AP (2009) Damage Detection in Repairs Using Frequency Response Techniques. Compos Struct 87:175-181

White L, Deleersnijder E, Legat V (2008) A Three-dimensional Unstructured Mesh Finite Element Shallow-water Model, with Application to the Flows around an Island and in a Wind-driven, Elongated Basin. Ocean Model 22:26–47

Wong JM, Mackie K, Stojadinovic B (2006) Bayesian Updating of Bridge Fragility Curves Using Sensor Data. In: Proc of the 3rd Intl Conf on Bridge Maint, Saf and Manag:613–614

Wu JE, Dai H (2008) Regularized Lanczos Method for Model Updating. J of Vib and Shock 27:65–69

Yan GR, Duan ZD, Ou JP (2007) Application of Genetic Algorithm on Structural Finite Element Model Updating. Harbin J of Harbin Inst of Technol 39:181–186

Yang Z, Wang L, Li B, Liu J (2009) Objective Functions and Algorithms in Structural Dynamic Finite Element Model Updating. Chin J of Appl Mech 26:288–296

Yao CR, Li YD (2008) Updating of Cable-stayed Bridges Model Based on Static and Dynamic Test Data. J of the Chin Railw Soc 30:65–70

Yeung DS, Chan PPK, Ng WWY (2009) Radial Basis Function Network Learning Using Localized Generalization Error Bound. Inf Sci 179:3199–3217

Yuan XX, Mao D, Pandey MD (2009) A Bayesian Approach to Modeling and Predicting Pitting Flaws in Steam Generator Tubes. Reliab Eng and Syst Saf 94:1838–1847

Yuan Y, Dai H (2006) Updating Finite Element Analytical Models Using Incomplete Modal Data Measured. J of Vib and Shock 25:154–156

Yuan Y, Dai H (2009) The Direct Updating of Damping and Gyroscopic Matrices. J of Comput and Appl Math 231:255–261

Zapico JL, Gonzlez-Buelga A, Gonzlez MP, Alonso R (2008) Finite Element Model Updating of a Small Steel Frame Using Neural Networks. Smart Mater and Struct 17:Art No 045016

Zárate BA, Caicedo JM (2008) Finite Element Model Updating: Multiple Alternatives. Eng Struct 30:3724–3730

Zhang LZ, Huang Q (2008) Updating of Bridge Finite Element Model Based on Optimization Design Theory. J of Harbin Inst of Technol 40:246–249

Zhang QH, Teo EC (2008) Finite Element Application in Implant Research for Treatment of Lumbar Degenerative Disc Disease. Med Eng and Phys 30:1246–1256

Zheng YM, Sun HH, Zhao X, Chen W, Zhang RH, Shen XD (2009) Finite Element Model Updating of a Long-span Steel Skybridge. J of Vib Eng 22:105–110

Zhong S, Oyadiji SO (2008) Analytical Predictions of Natural Frequencies of Cracked Simply Supported Beams with a Stationary Roving Mass. J of Sound and Vib 311:328–352

Zhong S, Oyadiji SO, Ding K (2008) Response-only Method for Damage Detection of Beam-like Structures Using High Accuracy Frequencies with Auxiliary Mass Spatial Probing. J of Sound and Vib 311:1075–1099

Zienkiewicz OC (1986) The Finite Element Method. McGraw-Hill Companies, New York

Zimmerman DC, Kaouk M (1992) Eigenstructure Assignment Approach for Structural Damage Detection. Am Inst of Aeronaut and Astronaut J 30:1848–1855

Zimmerman DC, Kaouk M, Simmermacher T (1995) Structural Damage Using Frequency Response Functions. In: Proc of the 13th Intl Modal Anal Conf:179–184

Živanović S, Pavic A, Reynolds P (2007) Finite Element Modelling and Updating of a Lively Footbridge: The Complete Process. J of Sound and Vib 301:126–145

Zong ZH, Xia ZH (2008) Finite Element Model Updating Method of Bridge Combined Modal Flexibility and Static Displacement. Chin J of Highw and Transp 21:43–49

Chapter 2

Finite-element-model Updating Using Nelder–Mead Simplex and BFGS Methods

Abstract. This chapter presents the Nelder–Mead simplex method and the Broyden–Fletcher–Goldfarb–Shanno (BFGS) method for finite-element-model updating. The methods presented have been tested on a simple beam and an unsymmetrical H-shaped structure. It was noted that, on average, the Nelder–Mead simplex method gives more accurate results than did the BFGS method. This is mainly because the BFGS method requires the calculation of gradients, which is prone to numerical errors within the context of finite-element-model updating.

Keywords: Nelder–Mead, objective function, Broyden–Fletcher–Goldfarb–Shanno, reduction methods, expansion methods, Guyan reduction method

2.1 Introduction

In Chapter 1, the concept of finite-element-model updating was introduced. It was noted there that a model-updating process is essentially an optimization problem where the updating parameters are those parameters in the finite-element model that are deemed to be highly uncertain, while the objective function is some measure of the distance between the finite-element's predicted data and measured data.

The purpose of this chapter is to introduce the concept of the updating of finite-element models using the Nelder–Mead (NM) simplex and Broyden–Fletcher–Goldfarb–Shanno (BFGS) optimization methods. The NM simplex method was used because it was found to be advantageous in that it (Olsson and Nelson, 1975):

- gives good results in the early stages of the simulations;
- does not require the use of the gradient and the Hessian of the objective function;
- is computationally efficient; and
- is relatively simple to understand and use.

The BFGS method was used because it has the following advantages. It preserves the necessary conditions for convergence – for example stability, descending direction, and positive definiteness (Azizi *et al.*, 2005). There is always a mismatch between the coordinates of the measured data and those of the finite-element model (Friswell and Mottershead, 1995). Because of this mismatch, it is always a good idea to reduce the system's matrices so that they correspond to the measured degrees of freedom, or to expand the measured coordinates so that they are of the same size as those from the finite-element model.

The purpose of this chapter is also to review reductions and expansions methods that equalize the size of the measured data and the system matrices of the structure in question. One question that is of great importance is: how do we compare the measurements and data from the finite-element model? This chapter reports an investigation that was carried out on how the data can be compared to establish whether the finite-element model has been sufficiently well updated.

The presented NM and BFGS optimization methods were then used for finite-element updating and were tested on a simple beam and an unsymmetrical H-shaped structure.

2.2 Introduction to Structural Dynamics

In this chapter, the modal properties *i.e.,* natural frequencies and mode shapes were used as a basis for finite-element-model updating. Thus, these parameters are described in this section. The modal properties are related to the physical properties of the structure. All elastic structures may be described in terms of their distributed mass, damping and stiffness matrices in the time domain through the following expression (Paz and Leigh, 2003):

$$[M]\{\ddot{X}\}+[C]\{\dot{X}\}+[K]\{X\}=\{F\} \tag{2.1}$$

where $[M]$ is the mass matrix, $[C]$ is the damping matrix, $[K]$ is the stiffness matrix, $\{X\}$ is the displacement vector, $\{\dot{X}\}$ is the velocity vector, $\{\ddot{X}\}$ is the acceleration vector, and $\{F\}$ is the applied force vector.

If Equation 2.1 is transformed into the modal domain to form an eigenvalue equation for the *i*th mode, then (Ewins, 1995):

$$(-\overline{\omega}_i^2[M]+j\overline{\omega}_i[C]+[K])\{\overline{\phi}\}_i=\{0\} \tag{2.2}$$

Here, $j=\sqrt{-1}$; $\overline{\omega}_i$ is the *i*th complex eigenvalue; and $\{0\}$ is the null vector. In Equation 2.2 the real part of $\{\overline{\phi}\}_i$ corresponds to the normalized mode shape $\{\phi\}_i$ while the imaginary part of $\overline{\omega}_i$ corresponds to the natural frequency ω_i.

From Equation 2.2 it may be deduced that the changes in the mass and stiffness matrices cause changes in the modal properties of the structure.

Therefore, the modal properties can be identified through the identification of the correct mass and stiffness matrices. The frequency-response function (FRF) is defined as the ratio of the Fourier-transformed response to the Fourier-transformed force. The FRF may be written in terms of the modal properties by using the modal summation equation as follows (Fu and He, 2001):

$$H_{kl}(\omega) = \sum_{i=1}^{N} \frac{-\omega^2 \phi_k^i \phi_l^i}{-\omega^2 + 2\zeta_i \omega_i \omega j + \omega_i^2} \tag{2.3}$$

In Equation 2.3 $H_{kl}(\omega)$ is an FRF due to excitation at position k and response measurement at position l, ω is the frequency point, ω_i is the ith natural frequency point, N is the number of modes and ζ_i is the damping ratio of mode i.

The excitation and response of the structure and Fourier-transform method (Ewins, 1995) can be used to calculate the FRF. Through Equation 2.3 and modal analysis (Ewins, 1995; Fu and He, 2001), the natural frequencies and mode shapes can be indirectly calculated from the FRFs. The modal properties of a dynamic system depend on the mass and stiffness matrices of the system as indicated by Equation 2.3. Therefore, the measured modal properties can be reproduced by the model if the correct mass and stiffness matrices are identified.

The finite-element-model updating process is achieved by identifying the correct mass and stiffness matrices. In the light of the measured data, the correct mass and stiffness matrices can be obtained by identifying the correct moduli of elasticity of various sections of the structure under consideration. In this chapter, to identify correctly the moduli of elasticity that would give the updated finite-element model, the following objective function that measures the distance between measured modal data and finite-element-model calculated modal data, was minimized (Marwala, 1997):

$$E = \left\| \sum_{i=1}^{N} \left[(-\omega_i^2 [M] + j\omega_i [C] + [K]) \{\phi\}_i \right] \right\| \tag{2.4}$$

Here, N is the number of measured modes; E is the error; and $\| \ \|$ is the Euclidean norm.

In Equation 2.4 the mass, damping and stiffness matrices are obtained from the finite-element model, while the natural frequencies and mode shapes are measured. If the natural frequencies and mode shapes of the system are described by the mass, damping and stiffness matrices then E is equal to zero. Therefore, the minimization of E identifies the updated finite-element model. Thus, the process of finite-element-model updating may be viewed as being an optimization problem. The updated finite-element models of a simple beam and an unsymmetrical H-shaped structure that are identified in this chapter were evaluated by comparing the natural frequencies and mode shapes from the finite-element models before and after updating to the measured ones.

2.3 Expansion/Reduction Methods

2.3.1 Model Expansion and Reduction Procedures

Two approaches may be pursued to ensure that the measured coordinates and modes are equal to the computed ones. These approaches are:

1. the experimental data may be expanded to the same number of degrees of freedom as the computed ones; and
2. the computed results may be reduced to the same number of coordinates as the measured ones.

Several techniques may be employed. As part of this study, the methods were implemented, and their effectiveness evaluated with respect to each method. The reduction methods applied were (Friswell and Mottershead, 1995):

- Guyan static reduction method;
- Guyan dynamic reduction method;
- Improved reduced system (IRS); and
- System equivalent reduction expansion process (SEREP).

The expansion methods applied were:

- expansion using mass and stiffness matrices; and
- expansion using modal data.

2.3.2 Model Reduction

A) Guyan Static Reduction (GSR) Method

This method was used to reduce the mass and stiffness matrices to the levels of the measured degrees of freedom. Prakash and Prabhu (1986) used the GSR method for reducing the mass and stiffness matrices in dynamic substructures, whereas Bushard (1981) applied this method to reduce system matrices in thermal problems. Bouhaddi and Fillod (1992) applied the GSR method for choosing master degrees of freedom in substructuring, while Häggblad and Eriksson (1993) applied the model condensation technique for analyses of large structures. Applications of reduction methods to nonlinear systems include Noor (1981).

In the GSR method, the state and force vectors, $\{x\}$ and $\{f\}$, and the mass and stiffness matrices $[M]$ and $[K]$ are partitioned into the measured (master) and unmeasured (slave) coordinates as follows (Friswell and Mottershead, 1995):

$$\begin{bmatrix} [M_{mm}][M_{ms}] \\ [M_{sm}]\ [M_{ss}] \end{bmatrix} \begin{Bmatrix} \ddot{x}_m \\ \ddot{x}_s \end{Bmatrix} + \begin{bmatrix} [K_{mm}][K_{ms}] \\ [K_{sm}]\ [K_{ss}] \end{bmatrix} \begin{Bmatrix} x_m \\ x_s \end{Bmatrix} = \begin{Bmatrix} f_m \\ 0 \end{Bmatrix} \tag{2.5}$$

Here, the subscripts m and s correspond to master and slave coordinates, respectively. The inertia terms are neglected to obtain the equation (Guyan, 1965):

$$[K_{sm}]\{x_m\} + [K_{ss}]\{x_s\} = [T_s]\{x_m\} \tag{2.6}$$

This equation may be used to eliminate the slave coordinate to leave the following equation (Guyan, 1965):

$$\begin{Bmatrix} x_m \\ x_s \end{Bmatrix} = \begin{bmatrix} [I] \\ -[K_{ss}]^{-1}[K_{sm}] \end{bmatrix} \{x_m\} = [T_s]\{x_m\} \tag{2.7}$$

The parameter T_s denotes the static transformation between full state vector and master coordinates and the parameter $[I]$ is the identity matrix. The reduced mass $[M_R]$ and stiffness $[K_R]$ matrices can be calculated as follows:

$$[M_R] = [T_s]^T [M][T_s] \tag{2.8}$$

and

$$[K_R] = [T_s]^T [K][T_s] \tag{2.9}$$

The frequency-response functions generated by the reduced-mass matrices are exact at zero frequency because the inertia matrices were neglected.

B) Guyan Dynamic Reduction (GDR) Method

The GSR method neglects the effects of inertia. However, the GDR method takes into account the inertia effect, assuming a particular frequency. Salvini and Vivio (2007) applied the GDR method to reduce the system matrices, while Yang (2009) applied it in model reduction using a Neumann series expansion. Yin *et al.* (2009) applied the GDR method to structural damage detection in a transmission tower. The technique used ambient vibration data.

The choice of frequency affects the accuracy of the reduced model. In the GDR method, the mass and stiffness matrices are partitioned into slave and master coordinates. The modified transformation matrices are as follows (Paz, 1984):

$$\begin{Bmatrix} \phi_m \\ \phi_s \end{Bmatrix} = \begin{bmatrix} [I] \\ -([K_{ss}]-\omega^2[M_{ss}])^{-1}([K_{sm}]-\omega^2[M_{sm}] \end{bmatrix}\{\phi_m\} = [T_D]\{\phi_m\} \quad (2.10)$$

The dynamic transformation, $[T_D]$, may then be used in the same way as the static transformation, $[T_S]$, to obtain the reduced mass and stiffness matrices similar to Equations 2.8 and 2.9.

C) Improved Reduced System (IRS)

The IRS is an improvement of the Guyan static reduction method. This method uses the transformation, the reduced mass and stiffness matrices from the Guyan reduction method together with the [S] matrix, which is made out of zeros and the inverse of the slave partition of the stiffness matrix, to obtain a new transformation matrix (O'Callahan, 1989):

$$[T] = \begin{bmatrix} [I] \\ -[K_{ss}]^{-1}[K_{sm}] \end{bmatrix} + [S][M][S][M_R]^{-1}[K_R] \quad (2.11)$$

where

$$[S] = \begin{bmatrix} [0][0] \\ [0][K_{ss}]^{-1} \end{bmatrix} \quad (2.12)$$

Friswell *et al.* (1995) studied the convergence of the iterated IRS method, whereas Kim and Cho (2008) studied the subdomain optimization of a multi-domain structure built by the IRS reduced system. Li *et al.* (2008) refined reduced models of dynamic systems, while Xia and Lin (2004) improved the iterated IRS method and applied this to structural eigensolutions. Friswell *et al.* (1998) developed iterated IRS techniques in structural dynamics. The iterated IRS method converges to the same transformation as the SEREP, which is the subject of the next section.

D) System Equivalent Reduction Expansion Process (SEREP)

The system equivalent reduction expansion process (SEREP) (O'Callahan *et al.*, 1989) partitions the analytical mode shapes into measured and unmeasured coordinates, and obtains the transformation by multiplying that with the generalized pseudo-inverse. See Equations 2.13 and 2.14.

$$[\phi]=\begin{bmatrix}\phi_m\\ \phi_s\end{bmatrix} \tag{2.13}$$

$$[T_U]=\{\phi\}\phi^+ \tag{2.14}$$

where

$$\phi^+=\left(\phi_m^T\phi_m\right)^{-1}\phi_m^T \tag{2.15}$$

The transformation may then be used in the same way as the static transformation, $[T_S]$, to obtain the reduced mass and stiffness matrices similar to Equations 2.8 and 2.9. Das and Dutt (2008) used the SEREP to reduce the model of a rotor-shaft system, while Sastry *et al.* (2003) introduced an iterative SEREP for extracting the high-frequency response.

2.3.3 Model Expansion

A) Expansion Using Mass and Stiffness Matrices (EMS)

In essence, this method is the inverse of the Guyan reduction method. Suppose ω_{mj} and ϕ_{mj} are the measured natural frequencies and mode shapes of coordinates i. Then the mass and the stiffness matrices from the finite-element analysis may be partitioned into measured and unmeasured coordinates. The equation of motion may then be written as follows (Friswell and Mottershead, 1995):

$$\left(-\omega_{mj}^2\begin{bmatrix}[M_{mm}][M_{ms}]\\ [M_{sm}][M_{ss}]\end{bmatrix}+\begin{bmatrix}[K_{mm}][K_{ms}]\\ [K_{sm}][K_{ss}]\end{bmatrix}\right)\begin{Bmatrix}\phi_{mj}\\ \phi_{sj}\end{Bmatrix}=\begin{Bmatrix}0\\ 0\end{Bmatrix} \tag{2.16}$$

where ϕ_{sj} represents the mode shape at the slave or the unmeasured coordinates. Rearranging the lower part of the matrix equation produces a solution for the unknown part of the measured mode shape vector. Thus (Friswell and Mottershead, 1995):

$$\{\phi_{sj}\}=-(-\omega_{mj}^2[M_{ss}]+[K_{ss}])^{-1}(-\omega_{mj}^2[M_{sm}]+[K_{sm}])\{\phi_{mj}\} \tag{2.17}$$

Other estimates of the unmeasured degrees of freedom may be obtained by using the upper part of Equation 2.12, which will involve the pseudo inverse. Using the upper part is satisfactory if the number of measured degrees of freedom exceeds the number of unmeasured degrees of freedom. Similarly, the unmeasured FRF may be calculated using the following equation (Friswell and Mottershead, 1995):

$$[H_{sj}] = -(-\omega_{mj}^2 [M_{ss}] + [K_{ss}])^{-1} (-\omega_{mj}^2 [M_{sm}] + [K_{sm}]) \{H_{mj}\} \quad (2.18)$$

Corus *et al.* (2006) applied this technique to improve structural dynamics models, while Kammer and Peck (2008) applied the expansion technique for sensor placement and Kammer (2005) applied this method for improved modal vibration testing.

B) Expansion Using Modal Data (EMD)

This method uses the modal data obtained from the finite-element model to estimate the modes at the unmeasured degrees of freedom. The measured modes are assumed to be a linear combination of the analytical modes at measured degrees of freedom and a transformation, *T*, as indicated by the following equation (Friswell and Mottershead, 1995):

$$\{\phi_m\} = [\phi_a]_m [T] \quad (2.19)$$

where $[\phi_a]_m$ represents the analytical mode shapes at measured degrees of freedom. Applying the pseudo-inverse to Equation 2.19 gives the transformation as (Friswell and Mottershead, 1995):

$$[T] = [\phi_a]_m^+ \{\phi_m\} \quad (2.20)$$

whcrc + indicates the pseudo-inverse. This transformation may be used to estimate the modes at unmeasured degrees of freedom from the finite-element analysis. It may also be used to smooth out the measured modes. Thus (Friswell and Mottershead, 1995):

$$\{\phi_m\}_{smoothed} = [\phi_a]_m [T] \quad (2.21)$$

$$[\phi_s] = [\phi_a]_m [T] \quad (2.22)$$

where $[\phi_a]_m$ represents the analytical mode shapes at the unmeasured degrees of freedom. The transformation may be obtained by using only the analytical modal data or the combination of measured and analytical data. This method is like an inverse of the SEREP method.

Marwala (1997) compared these methods and found that the reduction methods are more reliable than the expansion methods. This is because the expansion methods are more computationally intensive when compared to the reduction methods. Furthermore, the SEREP method was found to be susceptible to numerical instability. The IRS was found to be the best reduction method over the other methods.

2.4 Methods for Comparing Data

The most vital characteristic of modal testing is a comparison between the computed dynamic properties and those actually observed in practice (Ewins, 1995 Marwala, 1997; Ewins, 2001). This procedure is frequently called "validating" a theoretical model and it involves a number of stages. The first stage is to compare the specific dynamic properties, as measured against the predicted ones. The second stage is to measure the degree of the discrepancies or similarities between the two sets of data. The third stage is to bring the theoretical model closer to the measured data. When this is accomplished, the theoretical model is said to have been *updated.* In this section we closely study the computational methods used in the first, second and third stages.

In most situations, much endeavor goes into deriving the theory-based model and the experimentally derived model. For this reason, it is prudent to compare on as many different levels as possible. The dynamic model of a structure may be classified into *spatial, modal,* and *response* models (Ewins, 2001).

It is, at this time, appropriate to revisit this classification and attempt to compare the experimental and the theoretical model at each of these classifications. Consequently, a comparison of response properties as well as modal properties will be made. Comparisons between spatial properties are complex and, for that reason, will not be reflected on. In using any medium of comparison, the model must be developed comprehensively from the original form.

2.4.1 Direct Comparison

A) Comparisons of Natural Frequencies

The simplest method of comparison between the experimental and theoretical model is by comparing the natural frequencies (Ewins, 2001). This may be achieved by tabulating the experimental and theoretical natural frequencies. The most convenient way of comparison is to plot the graph of the experimental natural frequencies against the analytical ones for all available modes. If the gradient of the best straight line passing through the points is close to zero, then the correlation between the experimental and computed model is good. If the points lie spread out extensively about a straight line, then there is a severe failure of the model in representing the theoretical model's capacity to estimate the measured natural frequencies and, therefore, the theoretical model should be re-evaluated. If the positions diverge to some extent from the straight line, in a systematic fashion, this implies that there is a particular characteristic responsible for the deviation.

B) Comparisons of Mode Shapes

The mode shapes can also be compared by plotting the analytical modes against the experimental ones. For a simple structure with well-separated modes, this technique of comparison may be used with ease. Nevertheless, for a complicated structure with modes that are close to one another, this technique frequently becomes tricky to employ. Therefore, it is appropriate to make comparisons of mode shapes at the same time as those of the natural frequencies.

In the situation where we have more data to handle for each mode, the comparison may be conducted by plotting the deformed shape for each model, experimental and theoretical, and overlaying one plot on the other. The disadvantage of this technique is that, even though the difference is evident, the plots are not easy to understand and they are usually confusing because there is so much information. A suitable method of comparison, which is along the lines of the natural-frequency plot, is to plot each element in the mode shape vector, experimental and theoretical, on an *x-y* plot (Ewins, 2001).

The individual points on this graph relate to modal coordinates, and it is expected that they should lie close to a straight line. If the mode shape vectors are mass-normalized, this straight line should have a slope of 1. If the points lie close to a straight line with a slope that is not 1, then one of the mode shapes is not mass-normalized *or* there is scaling error in the data. If the points are widely spread out about the line, then there is inaccuracy in one set or the other set. If the dispersion is too great, then it may be that the eigenvectors that are being compared do not relate to the same mode.

The slope of the best straight line is called the modal scale factor (MSF) and is defined as (Ewins, 2001):

$$MSF\ (\phi_a, \phi_m) = \frac{\{\phi_a\}^T \{\phi_m\}^*}{\{\phi_a\}^T \{\phi_a\}^*} \tag{2.23}$$

Here, ϕ_a is the analytical modes; ϕ_m is the measured modes; and * is the complex conjugate.

The MSF parameter gives no indication of the quality of the measured points with respect to the straight-line fit.

2.4.2 Frequency-response Functions Assurance Criterion (FRFAC)

The benefit of using frequency-response functions (FRFs) straightforwardly is that they are measured directly. The easiest approach through which measured FRFs may be compared to the computed FRFs is by plotting the measured and theoretical FRFs in one plot (Ewins, 2001).

There are many measurements to be compared for particular FRF measurements. So, it becomes necessary to introduce a scalar factor (FRFAC) that gives the correlation between the measured and theoretical FRFs. One such scalar factor uses the measured FRFs directly (Marwala, 1997) and can be written as follows:

$$FRFAC = \frac{\sum_{j=1}^{M}\sum_{n=1}^{N}\left|H_a(n,j)\right|}{\sum_{j=1}^{M}\sum_{n=1}^{N}\left|H_m(n,j)\right|} \tag{2.24}$$

Here, N is the number of degrees of freedom; M is the number of the measured frequency; H_m is the measured FRF; and H_a is the analytical FRF.

An FRFAC of 1 indicates that the measured FRFs perfectly reflect the analytical FRFs; the FRFAC that is greater than 1 indicate that the magnitude of the analytical FRFs is, on average, greater than the experimental ones. An FRFAC that is less than 1 indicates that the magnitudes of the experimental FRFs are, on average, greater than the analytical ones.

2.4.3 The Model Assurance Criterion (MAC)

The modal assurance criterion (MAC) compares the measured and the computed mode shapes. Stetson (2008) calculated the MAC from electronic holography data, while Allemang (2003) reviewed the use of the modal assurance criterion over a period of 20 years. Yuan *et al.* (2009a) used the MAC optimally place a sensor on a cable-stayed bridge, whereas Caponero *et al.* (2002) used an interferometer and the MAC for identification of component modes. Brechlin *et al.* (1998) introduced the scaled modal assurance criterion to analyze a system with rotational degrees of freedom, whereas Lars (1998) used the modal assurance criteria to analyze two orthogonal modal vectors. Finally, Desforges *et al.* (1996) used the MAC for tracking modes during flutter testing while Heylen and Janter (1989) applied the modal assurance criterion for dynamic model updating. The MAC can be mathematically summarized by the following equation (Allemang and Brown, 1982):

$$MAC_{cdr} = \frac{\left|\{\phi_{cr}\}\{\phi_{dr}^*\}\right|^2}{\{\phi_{cr}\}^T\{\phi_{cr}^*\}\{\phi_{dr}\}^T\{\phi_{dr}^*\}} \tag{2.25}$$

Here, MAC is modal assurance criterion; c is for reference; d is the degrees-of-freedom; r is the mode; T is the transpose; * is the complex conjugate; and { } is a vector.

The MAC is a measure of the least-squares deviation of the points from a straight-line correlation. A value close to 1 suggests that the two mode shapes are well correlated, while a value close to 0 indicates that the mode shapes are not correlated.

2.4.4 The Coordinate Modal Assurance Criterion (COMAC)

The COMAC technique is based on the same principle as the MAC and is, in essence, a measure of the correlation between the measured and the computed mode shapes for a given common coordinate. Meo and Zumpano (2008) used the COMAC for damage estimation on plate structures, while Zhao and DeWolf (2007) applied the COMAC for damage detection in a cracked I-shaped steel beam. The COMAC for coordinate *j* is given by (Lieven and Ewins, 1988):

$$COMAC(j) = \frac{\left(\sum_{r=1}^{L} \left| {}^{j}\phi_{ar} \, {}^{j}\phi_{mr}^{*} \right| \right)^2}{\sum_{r=1}^{L} ({}^{j}\phi_{ar})^2 \sum_{r=1}^{L} ({}^{j}\phi_{mr}^{*})^2} \tag{2.26}$$

Unlike the MAC, the COMAC does not have any difficulty in comparing modes that are close in frequency or that are measured at insufficient transducer locations. L is the total number of well-correlated modes as indicated by the MAC. A value close to 1 suggests a good correlation. If the mode shape vectors are used then the COMAC becomes a vector. In this chapter, the MAC and direct comparison of mode natural frequencies are used to evaluate the effectiveness of the finite-element-model updating.

2.5 Optimization Methods

2.5.1 Nelder–Mead Simplex Method

The Nelder–Mead (NM) simplex method is one of the most used, direct optimization methods. First, the algorithm generates a simplex having N+1 vertices (x_i) in an N-dimensional space. Zhao *et al.* (2009) applied a modified NM simplex search for unconstrained optimization. The results obtained showed that the modified technique performs better than the original NM optimization method. Coelho and Araujo (2009) applied the NM simplex method for the successful identification in a Hénon chaotic map, while Ouria and Toufigh (2009) applied it for solving unconfined seepage problems.

Kalantar and Zimmer (2009) used the NM simplex method for optimally localizing vehicle formations, whereas Mastorakis (2009) applied this technique, finite-element modeling and a genetic algorithm for solving the Schrodinger–Maxwell's equations. Jung and Kim (2009) applied a hybrid of genetic algorithm and the NM simplex method for finite-element updating in a numerical bridge model.

A simplex of a specific dimension (a) is initialized around x_o using the following rule (Luersen and Riche, 2004; Ransome, 2006):

$$x_i = x_0 + \frac{a}{n\sqrt{2}}\left(\sqrt{n+1}+n-1\right)e_i + \sum_{k=1\neq i}^{n}\left[\frac{a}{n\sqrt{2}}\left(\sqrt{n+1}-1\right)\right]e_k, i = 1, n \quad (2.27)$$

Here, e is the unit base vector.

The simplex vertices coordinates are changed through using the reflection, expansion and contraction operators. The procedure is explained as follows (Ransome, 2006):

- for a given iteration in the optimization procedure the vertex with the worst fitness measure as defined by Equation 2.4 is substituted by a new vertex;
- the coordinates of the new vertex are established by reflecting the old vertex's point about the outstanding vertices. An easy reflection of a two-dimensional simplex is illustrated in Figure 2.1;
- if the fitness measure of the current vertices is lower than the preceding removed vertex's fitness, the dimensions of the simplex are minimized and if not, it is enlarged; and
- this process is continued until the functional evaluation values of the vertices converge.

The convergence process is measured using the following inequality:

$$\sqrt{\sum_{i=1}^{n+1}\frac{(f_i-\bar{f})^2}{n}} < \tau \quad (2.28)$$

where τ is a small positive scalar and is computed by using:

$$\bar{f} = \frac{1}{n+1}\sum_{i=1}^{n+1} f_i \quad (2.29)$$

The advantages of the NM simplex method are (Bürmen *et al.*, 2006):

- it normally yields major progress in the initial iterations and rapidly gives acceptable results;
- it usually involves a single or double functional evaluation per iteration, apart from a shrink transformation;
- it succeeds in getting an excellent decrease in the function value by using comparatively few function evaluations; and
- it is simple to comprehend and use.

The disadvantages of the NM simplex are (McKinnon, 1998):

- it lacks a convergence theory;
- it experiences numerical failure even for smooth functions;

- it takes many iterations with insignificant improvement in the function value even when it is far from a minimum; and
- it experiences untimely convergence.

2.5.2 Quasi-Newton Broyden–Fletcher–Goldfarb–Shanno (BFGS) Algorithm

The quasi-Newton optimization method is a successful, robust and quadratically convergent optimization method that uses a gradient. The technique is a derivation of the Newton–Raphson method, with the difference being that the inverse of the second derivative is updated through using a one-dimensional or multi-dimensional Hessian estimation method. Ghosal and Chaki (2009) used the quasi-Newton method for estimating and optimizing the depth of penetration in CO_2 Laser-MIG welding. The Newton–Raphson algorithm is mathematically represented as follows (Ransome, 2006):

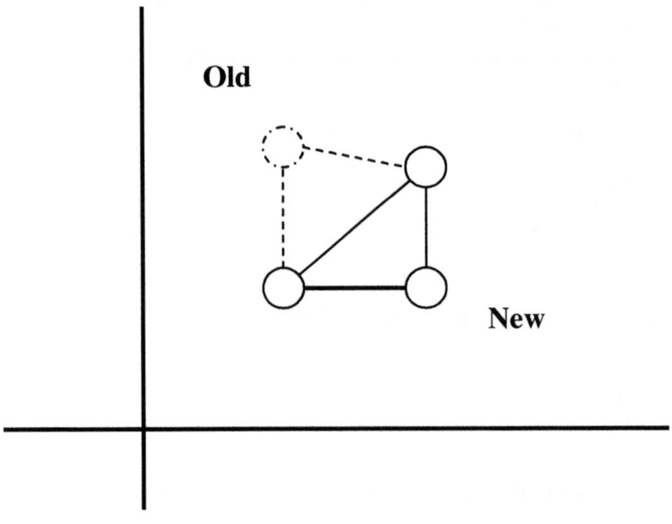

Figure 2.1 An easy reflection of a two-dimensional Simplex (Ransome, 2006)

$$x_{n+1} = x_n - \eta \frac{\dot{f}(x_n)}{\ddot{f}(x_n)} \tag{2.30}$$

Here, $f(x_n)$ is the objective function; and $\dot{f}(x_n)$ is the Jacobian (first-derivative), and $\ddot{f}(x_n)$ is the Hessian (second-derivative).

For a single-dimensional function, $\ddot{f}(x_n)$ can be updated by using the following equation:

$$\ddot{f}(x_{n+1}) = {}^{s_n}\!/_{y_n} \tag{2.31}$$

where

$$s_n = x_{n+1} - x_n \tag{2.32}$$

$$q_n = \dot{f}(x_{n+1}) - \dot{f}(x_n) \tag{2.33}$$

The most popular multi-dimensional technique for estimating the Hessian is the Broyden–Fletcher–Goldfarb–Shanno (BFGS) technique (Broyden, 1970; Fletcher, 1970; Goldfarb, 1970; Shanno, 1970). Sun *et al.* (2009) applied the BFGS technique for optimizing a machining allowance. Tan *et al.* (2009) applied the BFGS method for Stokes flows with fixed or moving interfaces and rigid boundaries, while Du *et al.* (2009) applied the BFGS method for optimizing the distribution of fibrous insulation. Further applications of the BFGS method include studies in aerodynamics (Papadimitriou and Giannakoglou, 2009), in solving equations (Yuan *et al.*, 2009b) and in nonconvex problems (Xiao *et al.*, 2009). The BFGS estimation of the inverse Hessian H_{n+1} is given by:

$$H_{n+1} = H_n + \frac{q_n q_n^T}{q_n^T s_n} - \frac{H_n^T s_n^T s_n H_n}{s_n^T H_n s_n} \tag{2.34}$$

From an initially estimated x_0 and Hessian matrix, H_0, repeat the following steps (Nocedal and Wright, 2006):

1. Obtain a step s_n by solving $s_n = -H_n^{-1}\nabla f(x_n)$ and do a line search to discover an optimal step size λ_n in the direction obtained in the initial step, and then update $x_{n+1} = x_n + \lambda_n s_n$;
2. $q_n = \nabla f(x_{n+1}) - \nabla f(x_n)$;
3. Calculate the $H_{n+1} = H_n + \frac{q_n q_n^T}{q_n^T s_n} - \frac{H_n^T s_n^T s_n H_n}{s_n^T H_n s_n}$.

2.6 Example 1: Simple Beam

The aluminum beam shown in Figure 2.2 was used to test the NM simplex method and the BFGS method for finite-element-model updating. This beam had the following dimensions: length: 1.1 m, width: 29.2 mm and thickness: 9.6 mm. This beam had holes of diameters 5.8 mm located at the centers of elements 2 to 9 and, therefore, was difficult to model. Further details of this beam are reported by Marwala (1997). The beam was tested freely suspended using elastic rubber bands. The beam was excited using an electromagnetic shaker and the response was measured using an accelerometer. The beam was also modeled using the structural dynamics toolbox (Balmès, 1997) and it was divided into 11 elements. The finite-element model used Euler–Bernoulli beam elements (Zienkiewicz, 1971). It was excited at a position indicated by the double arrows and acceleration was measured at 10 positions indicated by the single arrows in Figure 2.2. A set of 10 frequency response functions were calculated and a roving accelerometer was used for testing. The moduli of elasticity of these elements were used as updating parameters. When the finite-element-model updating was implemented, the moduli of elasticity were restricted to vary from 6×10^{10} to 8×10^{10} N m^{-2}. The NM simplex optimization method and the BFGS were run to optimize Equation 2.4 and the results in Table 2.1 were obtained.

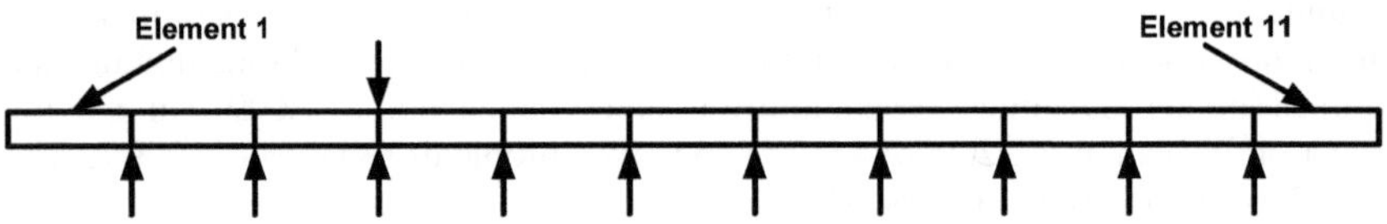

Figure 2.2 A beam with holes that is tested freely suspended

Table 2.1 Results showing measured frequencies of the beam, the initial frequencies and the frequencies obtained when the finite-element model of a beam is updated using the NM and BFGS

Modes	Measured frequency (Hz)	Initial frequency (Hz)	Frequencies from the NM updated model (Hz)	Frequencies from the BFGS updated model (Hz)
1	41.5	42.3	40.1	42.1
2	114.5	117.0	114.9	113.4
3	224.5	227.3	222.0	219.7
4	371.6	376.9	371.1	373.9

When a modulus of elasticity of 7×10^{10} N m^{-2} was assumed, the error between the first measured natural frequency and that from the initial finite-element model was 1.9%. When the NM was used for finite-element-model updating, this error was increased to 3.4% but using BFGS reduced it to only 1.4%.

The error between the second measured natural frequency and that from the initial model was 2.2%. When the NM was used, this error was reduced to 0.3% while using BFGS reduced it to 1.0%.

The error of the third natural frequency between the measured data and the initial finite-element model was 1.2%. When the NM was used, this error was reduced to 1.1%, while using the BFGS reduced it to 2.1%.

The error between the fourth measured natural frequency and that from the initial model was 1.4%. When the NM was used for finite-element-model updating, this error was reduced to 0.1%, while using the BFGS reduced it to 0.6%. Overall, the NM gave the best results.

The updated models were also validated on the mode shapes they predicted. To make this assessment possible, the modal assurance criterion (MAC) was used (Allemang and Brown, 1983). The mean of the diagonal of the MAC vector was used to compare the mode shapes predicted by the updated and initial finite-element models to the measured mode shapes. The average MAC calculated between the mode shapes from an initial finite-element model and the measured mode shapes was 0.9986. When the average MAC was calculated between the measured data and data obtained from the updated finite-element models, the results in Table 2.2 were obtained.

Table 2.2 Beam results showing the MAC calculated between measured mode shapes and the initial finite-element model for the NM and BFGS updated finite-element model

Method	MAC
Initial model	0.9986
NM	0.9988
BFGS	0.9986

It is observed that only the NM updated finite model gave improved averages for the diagonals of the MAC matrix. The computational time taken to run the complete NM method was 20 CPU sec, while the BFGS took 21 CPU min to run. In conclusion, the NM method was found to be better than the BFGS method. This is mainly because the implementation of the BFGS method experienced many numerical problems due to the difficulty in estimating the gradient and the Hessian matrix. This, in a way, suggests that in implementing computational methods, gradient-based methods should be avoided for finite-element-updating methods.

2.7 Example 2: Unsymmetrical H-shaped Structure

The unsymmetrical H-shaped aluminum structure shown in Figure 2.3 was also used to validate the proposed method (Marwala, 1997). This structure had three thin cuts of 1 mm that went half-way through the cross-section of the beam. These cuts were introduced to elements 3, 4 and 5. The structure with these cuts was used so that the initial FE model gave data that were far from the measured data and, thereby, tested the proposed procedure with a difficult finite-element-model updating problem. The structure was suspended using elastic rubber bands. It was excited using an electromagnetic shaker and the response was measured using an accelerometer. The structure was divided into 12 elements. It was excited at the position indicated by the double arrows in Figure 2.3, and the acceleration was measured at 15 positions indicated by the single arrows in Figure 2.3. The structure was tested freely suspended, and a set of 15 frequency-response functions were calculated. A roving accelerometer was used for measuring the response.

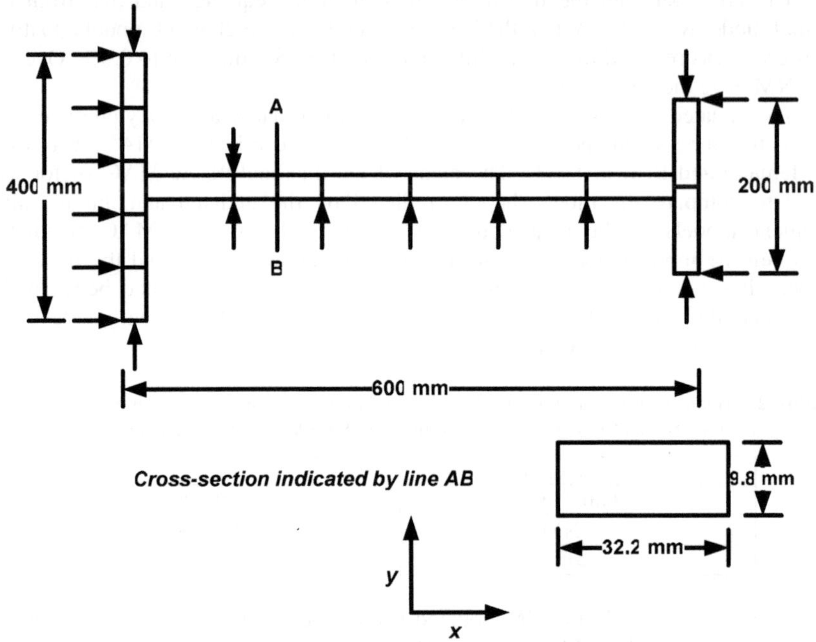

Figure 2.3 The irregular H-shaped structure

The mass of the accelerometer was found to be negligible compared to the mass of the structure. As in the previous example, the finite-element model was constructed using the structural dynamics toolbox (Balmès, 1997) with the Euler–Bernoulli beam elements (Zienkiewicz, 1971). The finite-element model contained 12 elements. The finite-element model of the structure contained 12 elements. As in the previous example, the moduli of elasticity of these elements were used as updating parameters, which were restricted to fall in the interval 6×10^{10} to 8×10^{10} N m^{-2}. The NM and BFGS were implemented as in the previous example.

These optimization methods were implemented for finite-element-model updating, and the results obtained are shown in Tables 2.3 and 2.4. Table 2.3 shows the measured natural frequencies, initial natural frequencies and natural frequencies obtained by the NM and BFGS updated finite-element models. When a modulus of elasticity of 7×10^{10} N m^{-2} was assumed, the error between the first measured natural frequency and that from the initial finite-element model was 4.3%. When the NM was used for finite-element-model updating, this error was reduced to 1.1% and the BFGS approach reduced this error to 2.6%.

The error between the second measured natural frequency and that from the initial model was 8.4%. When the NM was used, this error was reduced to 1.4% and the BFGS reduced this error to 5.8%.

The error of the third natural frequencies between the measured data and the initial finite-element model was 9.6%. When the NM was used, this error was reduced to 2.4% and in using the BFGS it was reduced to 1.2%.

The error between the fourth measured natural frequency and that from the initial model was 3.7%. When the NM was used, this error was reduced to 0.7% and using the BFGS it was reduced to 2.1%.

The error between the fifth measured natural frequency and that from the initial model was 1.6%. When the NM was used this error was reduced to 2.6% and the BFGS reduced it to 1.1%.

Overall, the NM gave the best results with an average error calculated over all the five natural frequencies of 2.1%, while the BFGS gave an average error of 2.6%. On average, all four methods improved when compared to the average error between the initial finite-element model and the natural frequencies, which was 5.5%.

Table 2.3 Results from an unsymmetrical H-shaped structure showing measured frequencies, the initial frequencies and the frequencies obtained when the finite-element model was updated using the NM and BFGS

Modes	Measured frequency (Hz)	Initial frequency (Hz)	Frequencies from the NM updated model (Hz)	Frequencies from the BFGS updated model (Hz)
1	53.9	56.2	53.3	55.3
2	117.3	127.1	118.9	124.1
3	208.4	228.4	213.3	205.8
4	254.0	263.4	252.3	248.6
5	445.1	452.4	433.7	440.4

As in the previous example, the updated models implemented were validated on the mode shapes they predicted using the MAC. The NM- and BFGS-updated finite-element models gave improved averages for the diagonals of the MAC matrices of 0.8404 and 0.8403, respectively. When the computational load for each method was monitored, it was observed that, on average, the NM was the most computationally efficient method.

Table 2.4 Beam results showing the MAC calculated between measured mode shapes and the initial finite-element model for the NM and BFGS updated finite-element model

Method	MAC
Initial model	0.8394
NM	0.8404
BFGS	0.8403

Again, the results indicate that the NM method is marginally better than the BFGS method. The reason for this seems to be the fact that the calculations of the gradient and the Hessian matrix were fraught with numerical difficulties.

2.8 Conclusion

In this chapter the NM simplex and the BFGS methods were implemented for finite-element-model updating. When these techniques were tested on a simple beam and an unsymmetrical H-shaped structure, it was observed, on average, that the NM simplex method gave more accurate results than the BFGS method. This is mainly because the BFGS methods require the calculation of gradients, which is fraught with numerical errors. Furthermore, the NM method was found to be more computationally efficient than the BFGS method. For further work, hybrid techniques should be explored.

2.9 Further Work

This chapter compared the Nelder–Mead simplex and the BFGS methods for finite-element-model updating. These methods are classified as local search methods. For future work, other local search methods such as conjugate gradient method and the Levenberg–Marquardt algorithm need to be implemented for finite-element-model updating. In particular, their convergence properties within the context of finite-element-model updating also need to be studied. Further work should also include hybridizing these local search methods.

References

Allemang RJ (2003) The Modal Assurance Criterion – Twenty Years of Use and Abuse. Sound and Vib 37:14–21

Allemang RJ, Brown DL (1982) A Correlation Co-efficient for Modal Vector Analysis. In:Proc of the 1st Int Modal Anal Conf:01–18

Azizi SM, Fekri MZ, Yazdani D, Bakhshai (2005), A. A BFGS-Based Method for Designing a Decentralized LQ Controller for Industrial Robot Manipulators. In:Proc of the IEEE Conf on Control Appl: 1558–1562

Balmès E (1997) Structural Dynamics Toolbox User's Manual. Sci Softw Group Sèvres Fr Ver 2.1

Bouhaddi N, Fillod R (1992) A Method for Selecting Master DOF in Dynamic Substructuring Using the Guyan Condensation Method. Comput & Struct 45: 941–946

Brechlin E, Bendel K, Keiper W (1998) New Scaled Modal Assurance Criterion for Eigenmodes Containing Rotational Degrees of Freedom. In: Proc of the 23rd Intl Conf on Vib Eng ISMA:465–471

Broyden CG (1970) The Convergence of a Class of Double-rank Minimization Algorithms. J of the Inst of Math and its Appl 6:76–90

Bürmen A, Puhan J, Tuma T (2006) Grid Restrained Nelder-Mead Algorithm. Comput Optim Appl 34:359–375

Bushard LB (1981) The Value of Guyan Reduction in Dynamic Thermal Problems. Comp & Struct 13:525–531

Caponero MA, Paolozzi A, Peroni I (2002) Use of Speckle Interferometry and Modal Assurance Criterion for Identification of Component Modes. Opt and Lasers in Eng 37:355–367

Coelho LdS, Araujo E (2009) Identification of the Hénon Chaotic Map by Fuzzy Modeling and Nelder-Mead Simplex Method Chaos. Solitons and Fractals 41:2762–2772

Corus M, Balmès E, Nicolas E (2006) Using Model Reduction and Data Expansion Techniques to Improve SDM. Mech Res Commun 35:398–407

Das AS, Dutt JK (2008) Reduced Model of a Rotor-Shaft System Using Modified SEREP. Mech Res Commun 35:398–407

Desforges MJ, Cooper JE, Wright JR (1996) Mode Tracking During Flutter Testing Using Modal Assurance Criterion. In Proc of the Inst of Mech Eng J of Aerosp Eng 210:27–37

Du N, Fan J, Wu H, Sun W (2009) Optimal Porosity Distribution of Fibrous Insulation. Intl J of Heat and Mass Transf 52:4350–4357

Ewins DJ (1995) Modal Testing: Theory and Practice. Research Studies Press, Letchworth

Ewins DJ (2001) Modal Testing: Theory, Practice and Application Mechanical Engineering Research Studies Engineering Dynamics Ser 2, Wiley

Fletcher RA (1970) New Approach to Variable Metric Algorithms. Compt J. 13:317–322

Friswell M I, Garvey SD, Penny JET (1995) Model Reduction Using Dynamic and Iterated IRS Techniques. J of Sound and Vib 186:311–323

Friswell M I, Garvey SD, Penny JET (1998) The Convergence of the Iterated IRS Method. J of Sound and Vib 211:123–132

Friswell MI, Mottershead JE (1995) Finite Element Model Updating in Structural Dynamics. Kluwer Academic Publishers Group, Norwell

Fu ZF, He J (2001) Modal Analysis. Butterworth-Heinemann Publishers, Oxford

Ghosal S, Chaki S (2009) Estimation and Optimization of Depth of Penetration in Hybrid CO2 LASER-MIG Welding using ANN-optimization Hybrid Model. Intl J of Adv Manuf Technol:01–9

Guyan RJ (1965) Reduction of Stiffness and Mass Matrices. AIAA J 3:380

Goldfarb DA (1970) Family of Variable Metric Updates Derived by Variational Means. Math of Comput 24:23–26

Häggblad RJ, Eriksson L (1993) Model Reduction Methods for Dynamic Analyses of Large Structures. Comput & Struct 47:735–749

Heylen W, Janter T (1989) Applications of the Modal Assurance Criterion in Dynamic Model Updating. Am Soc of Mech Eng, Des Eng Division DE 18:289–294

Jung DS, Kim CY (2009) FE Model Updating Based on Hybrid Genetic Algorithm and its Verification on Numerical Bridge Model. Struc Eng and Mech 32:667–683

Kalantar S, Zimmer UR (2009) Optima Localization by Vehicle Formations Imitating the Nelder-Mead Simplex Algorithm. Auton Robots:01–22

Kammer DC (2005) Sensor Set Expansion for Modal Vibration Testing. Mech Syst and Signal Process 19:700–713

Kammer DC, Peck JA (2008) Mass-weighting Methods for Sensor Placement Using Sensor Set Expansion Techniques. Mech Syst and Signal Process 22:1515–1525

Kim H, Cho M (2008) Sub-domain Reduction Method in Non-matched Interface Problems. J of Mech Sci and Technol 22:203-212

Lars R (1998) Modal Assurance Criteria Value for Two Orthogonal Modal Vectors. In: Proc of the Intl Modal Anal Conf – IMAC 2:1320–1325

Li H, Zhang M, Hu SJ (2008) Refinement of Reduced-models for Dynamic Systems. Prog in Nat Sci 18:993–997

Lieven NAJ, Ewins DJ (1988) Spatial Correlation of Mode Shapes the Coordinate Modal Assurance Criterion (COMAC). In Proc 6th Intl Modal Anal Conf:690–695

Luersen M, Riche RL (2004) Globalized Nelder-Mead Method for Engineering Optimization. Comput & Struct 82:2251–2260

Marwala T (1997) A Multiple Criterion Updating Method for Damage Detection on Structures. Master's Thesis, University of Pretoria.

Mastorakis NE (2009) Solution of the Schrodinger-Maxwell Equations via Finite Elements and Genetic Algorithms with Nelder-Mead. WSEAS Trans on Math 8:169–176

McKinnon KIM (1998) Convergence of the Nelder-Mead Simplex Method to a Nonstationary Point. SIAM J Optim 9:148–158

Meo M, Zumpano G (2008) Damage Assessment on Plate-like Structures Using a Global-Local Optimization Approach. Optim and Eng 9:161–177

Nocedal J, Wright SJ (2006) Numerical Optimization.Springer-Verlag, Berlin

Noor AK (1981) Recent Advances in Reduction Methods for Nonlinear Problems. Comput & Struct 13:31–44

O'Callahan JC (1989) A Procedure for an Improved Reduced System (IRS) model. In: Proc of the 7th Intl Modal Anal Conf:29–37

O'Callahan JC, Avitabile P, Riemer R (1989) System Equivalent Reduction Expansion Process. In: Proc of the 7th Intl Modal Anal Conf:29–37

Olsson DM, Nelson LS (1975) The Nelder-Mead Simplex Procedure for Function Minimization. Technometrics 17:45–51

Ouria A, Toufigh MM (2009) Application of Nelder-Mead Simplex Method for Unconfined Seepage Problems. Appl Math Model 33:3589–3598

Papadimitriou DI, Giannakoglou KC (2009) The Continuous Direct-Adjoint Approach for Second Order Sensitivities in Viscous Aerodynamic Inverse Design Problems.Comput and Fluids 38:1539–1548

Paz M (1984) Dynamic Condensation. AIAA J 22:724–727

Paz M, Leigh W (2003) Structural Dynamics: Theory and Computation, Springer

Prakash BG, Prabhu MSS (1986) Reduction Techniques in Dynamic Substructures for Large Problems. Comput & Struct 22:539–552

Ransome TM (2006) Automatic Minimisation of Patient Setup Errors in Proton Beam Therapy. Master's Thesis, University of the Witwatersrand

Salvini P, Vivio F (2007) Dynamic Reduction Strategies to Extend Modal Analysis Approach at Higher Frequencies. Finite Elem in Anal and Des 43:931–940

Sastry CVS, Mahapatra DR, Gopalakrishnan S, Ramamurthy TS (2003) An Iterative System Equivalent Reduction Expansion Process for Extraction of High Frequency Response from Reduced Order Finite Element Model. Comput Methods in Appl Mech and Eng 192:1821–1840

Shanno DF (1970) Conditioning of Quasi-Newton Methods for Function Minimization. Math of Comput 24:647–656

Stetson KA (2008) Calculating Modal Assurance Criteria from Electronic Holography Data. Sound and Vib 42:06–11

Sun YW, Xu JT, Guo DM, Jia ZY (2009) A Unified Localization Approach for Machining Allowance Optimization of Complex Curved Surfaces. Precis Eng 33:516–523

Tan Z, Lim KM, Khoo BC (2009) An Immersed Interface Method for Stokes Flows with Fixed/Moving Interfaces and Rigid Boundaries. J of Comput Phys 228:6855–6881

Xia Y, Lin R (2004) Improvement on the Iterated IRS Method for Structural Eigensolutions. J of Sound and Vib 270:713–727

Xiao Y, Sun H, Wang Z (2009) A globally Convergent BFGS Method with Non-monotone Line Search for Non-convex Minimization. J of Comput and Appl Math 230:095–106

Yang QW (2009) Model Reduction by Neumann Series Expansion. Appl Math Modelling 33:4431– 4434

Yin T, Lam HF, Chow HM, Zhu HP (2009) Dynamic Reduction-Based Structural Damage Detection of Transmission Tower Utilizing Ambient Vibration Data. Eng Struct 31:2009–2019

Yuan A, Dai H, Sun D (2009a) Optimal Sensor Placement of Cable-Stayed Bridge Using Mixed Algorithm Based on Effective Independence and Modal Assurance Criterion Methods. J of Vib, Measurement and Diagn 29:U466

Yuan G, Lu X, Wei Z (2009b) BFGS Trust-Region Method for Symmetric Nonlinear Equations. J of Comput and Appl Math 230:44–58

Zhao J, DeWolf JT (2007) Modeling and Damage Detection for Cracked I-shaped Steel Beams. Struct Eng and Mech 25:131–146

Zhao QH, Urosević D, Mladenović N, Hansen P (2009) A Restarted and Modified Simplex Search for Unconstrained Optimization. Comput and Oper Res 36:3263–3271

Zienkiewicz OC (1971) The Finite Element Method in Engineering Science. McGraw-Hill, London

Chapter 3

Finite-element-model Updating Using a Genetic Algorithm

Abstract. This chapter implements a genetic algorithm for finite-element-model updating. This method was tested on a simple beam and an unsymmetrical H-shaped structure and compared to a method based on the Nelder–Mead simplex method. It was observed on average that the genetic algorithm method gives more accurate results for modal properties than does the Nelder–Mead (NM) simplex method.

Keywords: Nelder–Mead, a genetic algorithm, finite-element-model updating, crossover, mutation, reproduction

3.1 Introduction

Numerical-model updating is a process of tuning the numerical model so that it better reflects the observed data from the physical structure being modeled (Friswell and Mottershead, 1995). One such numerical model is the finite-element model. The finite-element model is an approximation to the true physical object being modeled and finite-element-model updating means identifying a better approximation model for the physical object than the original finite-element model. This process is fundamentally an optimization problem where the design variables are the parameters of the finite-element model that are deemed to be in doubt (and therefore need to be updated) and the objective function is an equation that characterizes some distance between the finite-element-model predictions and those from measurements.

Many optimization methods have been used for finite-element-model updating. Levin and Lieven (1998) introduced the genetic algorithm (GA) and simulated annealing for finite-element-model updating. The main drawback with the genetic algorithm is the speed at which the solution is arrived at, making it incredibly difficult to implement for a large-scale structure and furthermore, it contains many choices and parameters that need to be selected. Marwala (2002) successfully applied a genetic algorithm to minimize the distance between the measured wavelet data and the finite-element-predicted wavelet data. The drawback of this approach to finite-element model updating is the associated high computational expense of the genetic-algorithm simulation and the wavelet processing of the vibration data used. The advantages of a genetic algorithm over the gradient-based methods (Marwala and Heyns, 1998) are:

- the genetic algorithm has a higher probability of identifying a global optimum solution than the gradient-based approach; and
- finite-element-model updating produces a nonsmooth objective function that makes the process of calculating the gradient extremely difficult and sometimes impossible because of the occurrence of badly scaled matrices.

Any optimization method can be used for the finite-element-updating problem. Some of the successfully used methods include:

- the response-surface method that will be discussed in Chapter 6 (Marwala, 2004);
- particle-swarm optimization that is implemented in Chapter 4 and was successfully applied for optimizing the shape of structures to meet the design requirements (Fourie and Groenwold, 2000; Fourie and Groenwold, 2001) and then used subsequently for finite-element-model updating (Marwala, 2005); and
- simulated annealing, which is implemented in Chapter 5.

The aim of this chapter is to study the use of a genetic algorithm for finite-element-model updating. This procedure is then compared to the Nelder–Mead simplex method (Nelder and Mead, 1965).

Finite-element-model updating has been widely used to detect damage in structures (Doebling *et al.*, 1996). When implementing finite-element-model updating methods for damage identification, it is assumed that the finite-element model is an accurate dynamic representation of the structure and finite-element-model updating is used to achieve this accuracy. This means that changing any physical parameter of an element in the finite-element model is equivalent to introducing "damage" in that region.

As described before, there are two approaches used in finite-element-model updating: direct methods and iterative methods (Friswell and Mottershead, 1995). Direct methods, which use the modal properties, are computationally efficient to implement and reproduce the measured modal data exactly. Furthermore, they do not take into account the physical parameters that are updated. Consequently, even though the finite-element model can predict measured quantities, the updated model is limited in the following ways:

- it may lack connectivity of nodes – connectivity of nodes is a phenomenon that occurs naturally in finite-element modeling because of the reality that the structure is physical connected;
- the updated matrices are populated instead of banded – the fact that structural elements are only connected to their neighbors ensures that the

mass and stiffness matrices are diagonally dominated with few couplings between elements that are far apart; and

- there is a possible loss of symmetry of the systems' matrices.

Iterative procedures use changes in physical parameters to update finite-element models and produce models that are physically realistic. The advantages of iterative approaches to finite-element-model updating are that they:

- produce a finite-element model that can be interpreted physically, and
- can be implemented without the use of matrix reduction or expansion schemes as implemented in Chapter 2.

Iterative methods that use modal properties for finite-element-model updating and a genetic algorithm for optimization are implemented in this chapter. The finite-element models are updated so that the measured modal properties match the finite-element-model predicted modal properties. The finite-element-model updating procedures use the Nelder–Mead simplex method and the genetic algorithm for optimization and their performances are compared using a simple beam and an unsymmetrical H-shaped structure.

3.2 Mathematical Background

In this chapter, modal properties *i.e.*, natural frequencies and mode shapes, are used as a basis for finite-element-model updating. For this reason these parameters are described in this section. Modal properties are related to the physical properties of the structure. All elastic structures may be described in terms of their distributed mass, damping and stiffness matrices in the time domain through the following expression (Ewins, 1995):

$$[M]\{\ddot{x}\}+[C]\{\dot{x}\}+[K]\{X\}=\{F\} \tag{3.1}$$

Here, $[M]$ is the mass matrix; $[C]$ is the damping matrix; $[K]$ is the stiffness matrix; $\{x\}$ is a displacement vector; $\{\dot{x}\}$ is the velocity vector, $\{\ddot{x}\}$ is the acceleration vector; and $\{F\}$ is the applied force vector.

If Equation 3.1 is transformed into the modal domain to form an eigenvalue equation for the ith mode, then (Ewins, 1995):

$$(-\overline{\omega}_i^2[M]+j\overline{\omega}_i[C]+[K])\{\overline{\phi}\}_i=\{0\} \tag{3.2}$$

Here, $j=\sqrt{-1}$; $\overline{\omega}_i$ is the ith complex eigenvalue (the imaginary part that corresponds to the natural frequency ω_i); $\{0\}$ is the null vector; and $\{\overline{\phi}\}_i$ is the ith complex mode shape vector (the real part which corresponds to the mode shape $\{\phi\}_i$). From Equation 3.2, it may be deduced that the changes in the mass and stiffness matrices cause changes in the modal properties of the structure. The modal properties can, therefore, be identified through the identification of the correct mass and stiffness matrices.

The frequency-response function (FRF) is defined as the ratio of the Fourier-transformed response to the Fourier-transformed force (Ewins, 1995). The FRF (H) may be written in terms of the modal properties by using the modal summation equation as follows (Ewins, 1995):

$$H_{kl}(\omega) = \sum_{i=1}^{N} \frac{-\omega^2 \phi_k^i \phi_l^i}{-\omega^2 + 2\zeta_i \omega_i \omega j + \omega_i^2} \tag{3.3}$$

Here, $H_{kl}(\omega)$ is an FRF due to excitation at k and response measurement at l; ω is the frequency point; ω_i is the ith natural frequency point; N the number of modes; and ζ_i is the damping ratio of mode i.

The excitation and response of the structure and Fourier-transform method (Ewins, 1995) can be used to calculate the frequency-response-function. Through Equation 3.3 and modal analysis (Ewins, 1995), the natural frequencies and mode shapes can be indirectly calculated from the frequency-response functions. The modal properties of a dynamic system depend on the mass and stiffness matrices of the system as indicated by Equation 3.2.

The measured modal properties can therefore be reproduced by the model if the correct mass and stiffness matrices are identified. The finite-element-model updating is achieved by identifying the correct mass and stiffness matrices that ensure that the finite-element model predicts the measured data to the sufficient accuracy.

In the light of the measured data, the correct mass and stiffness matrices can be obtained by identifying the correct moduli of elasticity of various sections of the structure under consideration.

In this chapter, to correctly identify the moduli of elasticity that would give the updated finite-element model, the following objective function is minimized that measures the distance between measured modal data and finite-element-model calculated modal data:

$$E = \sum_{i=1}^{N} \gamma_i \left(\frac{\omega_i^m - \omega_i^{calc}}{\omega_i^m} \right)^2 + \beta \sum_{i=1}^{N} \left(1 - diag(MAC(\phi_i^{calc}, \phi_i^m))\right) \tag{3.4}$$

Here, M indicates a measured variable; calc indicates a calculated variable; ω_i is the ith natural frequency; ϕ_i is the ith mode shape vector; N is the number of modes; γ_i is the weighting factor that measures the relative distance between the initial estimated natural frequencies for mode i and the target frequency of the same mode; β is the weighting function on the mode shapes; MAC is the modal assurance criterion (Allemang and Brown, 1982); and $diag(MAC)_i$ stands for the ith diagonal element of the MAC matrix.

It should be noted that the objective function in Equation 3.4 is not the same as in Chapter 2 as it does not implicitly take into account the equation of motion, as was the case in Chapter 2. Conventionally, Equation 3.4 includes the regularization parameters (Friswell and Mottershead, 1995). Regularization is a process of introducing an additional penalty function to solve an ill-posed problem or prevent over-fitting the model (Tibshirani, 1996). In essence, this ensures that the finite-element model is not too tuned to reflect the measured data at the expense of the physics in the finite-element model. In this chapter, to deal with this potential problem, particular attention is paid to the choice of the bounds of the parameters to be updated, with an emphasis on the engineering judgment of the user.

In Equation 3.4, the first part has the function of ensuring that the natural frequencies predicted by the finite-element model are as close to the measured ones as possible, while the second term ensures that the mode shapes between measurements and finite-element model prediction are correlated. The MAC is a measure of the correlation between two sets of mode shapes of the same dimension and can be represented mathematically as follows (Allemang and Brown, 1982):

$$MAC_{cdr} = \frac{\left|\{\phi_{cr}\}\{\phi_{dr}^{*}\}\right|^{2}}{\{\phi_{cr}\}^{T}\{\phi_{cr}^{*}\}\{\phi_{dr}\}^{T}\{\phi_{dr}^{*}\}} \tag{3.5}$$

Here, MAC is the modal assurance criterion; C is for reference; d is the degree of freedom; r is the mode; T is the transpose; * is the complex conjugate; and { } is a vector.

The modal assurance criterion varies from zero to one, where zero represents no consistent correspondence and one represents a consistent correspondence. If the modal vectors under consideration really demonstrate a consistent, linear relationship, the modal assurance criterion ought to approach 1 (Allemang and Brown, 1982). The updated finite-element models of the simple beam and an unsymmetrical H-shaped structure that are identified in this chapter are evaluated by comparing the natural frequencies and mode shapes from the finite-element models before and after updating to the measured ones.

3.3 Genetic Algorithm

The finite-element-model updating method presented in this chapter uses a genetic algorithm. Unlike many optimization algorithms, a genetic algorithm method has a higher probability of converging to a global optimal solution than a gradient-based method. A genetic algorithm is a population-based, probabilistic technique that operates to find a solution to a problem from a population of possible solutions (Kubalik and Lazanský, 1999). It is used to find approximate solutions to difficult problems through the application of the principles of evolutionary biology to computer science (Michalewicz, 1996; Mitchell, 1996; Forrest, 1996; Vose, 1999; Tettey and Marwala, 2006). It is analogous to Darwin's theory of evolution where members of the population compete to survive and reproduce while the weaker ones die out. Each individual is assigned a fitness value according to how well it meets the objective of solving the problem and, in this chapter, to identify the optimal updated finite-element model. New and more evolutionary-fit individual solutions are produced during a cycle of generations, wherein selection and recombination operations take place, analogous to how gene transfer applies to the current individuals. This continues until a termination condition is met, after which the best individual thus far is considered to be the updated finite-element model. This chapter describes the use of a genetic algorithms to optimize Equation 3.4.

Genetic algorithms have successfully been applied for finite-element updating in the past. These applications include Marwala (2002), who used wavelet data to update finite-element models in structures, Akula and Ganguli (2003), who applied finite-element-model updating, based on genetic algorithms to helicopter rotor-blade design. Perera and Ruiz (2008) successfully applied a genetic-algorithm-based, finite-element-model updating for damage identification in large-scale structures, while Tu and Lu (2008) enhanced genetic-algorithm applications by considering artificial boundary conditions and Franulović *et al.* (2009) implemented a genetic algorithm for material model parameter identification for low-cycle fatigue.

The success of a genetic algorithm, for optimizing complex systems is not only limited to finite-element-model updating. Other recent successful applications include Balamurugan *et al.* (2008) who evaluated the performance of a two-stage adaptive a genetic algorithm, enhanced with island and adaptive features in structural topology optimization, while Kwak and Kim (2009) successfully implemented a hybrid genetic algorithm, enhanced by a direct search for optimum design of reinforced concrete frames. Canyurt *et al.* (2008) estimated the strength of a laser hybrid welded joint using a genetic-algorithm approach.

Perera *et al.* (2009) applied a genetic algorithm to assess the performance of a multi-criteria damage-identification system. Almeida and Awruch (2009) used a genetic-algorithm and finite-element models to optimally design composite laminated structures. The genetic algorithm was adapted with particular operators and variables codification for the definite class of composite laminated structures.

Further successful applications of a genetic algorithm for optimization structures include Paluch *et al.* (2008) as well as Roy and Chakraborty (2009). In addition, a genetic algorithm has also been proven to be very successful in many applications including:

- finite-element analysis (Marwala, 2003);
- selecting optimal neural-network architecture (Arifovic and Gençay, 2001);
- training hybrid fuzzy neural networks (Oh and Pedrycz, 2006);
- solving job-scheduling problems (Park *et al.*, 2003);
- remote sensing (Stern *et al.*, 2006);
- missing-data estimation (Abdella and Marwala, 2006); and
- combinatorial optimization (Zhang and Ishikawa, 2004).

Furthermore, the genetic-algorithm method has been proven to be successful in complex optimization problems such as wire routing, scheduling, adaptive control, game playing, cognitive modeling, transportation problems, traveling salesman problems, optimal control problems and database-query optimization (Pendharkar and Rodger, 1999; Marwala *et al.*, 2001; Marwala and Chakraverty, 2006; Marwala, 2007; Crossingham and Marwala, 2007; Hulley and Marwala, 2007). The MATLAB® implementation of a genetic algorithm described in Houck *et al.*

(1995) was used to implement the GA in this chapter. To implement a genetic algorithm, as shown in Figure 3.1, the following steps are followed: initialization, crossover, mutation, selection, reproduction and termination.

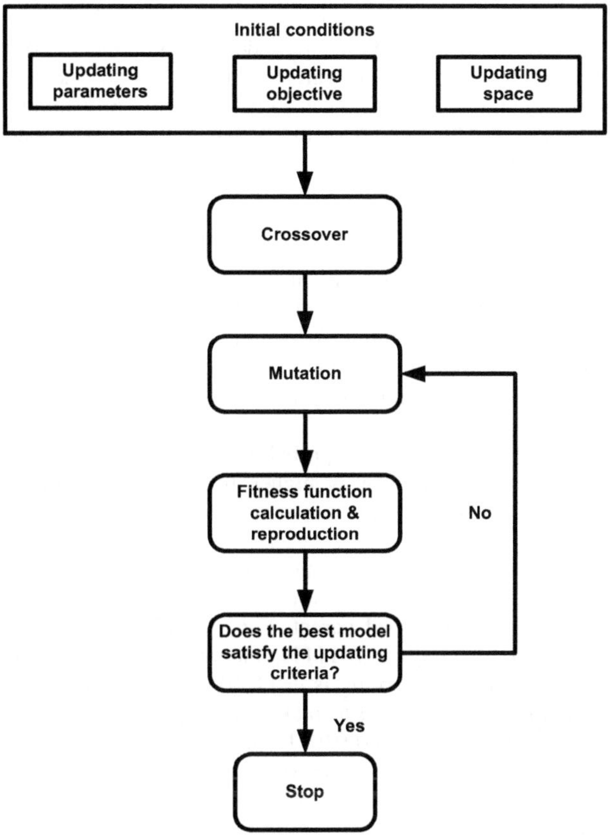

Figure 3.1 Flowchart of the genetic algorithm method

In this chapter, the genetic algorithm views learning as a competition among a population of evolving candidate problem solutions. A fitness function, which in this chapter is represented by Equation 3.4, evaluates each solution to decide whether it will contribute to the next generation of solutions. Through operations analogous to gene transfer in sexual reproduction, the algorithm creates a new population of candidate solutions (Goldberg, 1989). The three most important aspects of using a genetic algorithms are:

- the definition of the objective function;
- implementation of the genetic representation; and
- implementation of the genetic operators.

The details of genetic algorithms are illustrated in Figure 3.1.

3.3.1 Initialization

In the beginning, a large number of possible individual solutions are randomly generated to form an initial population. This initial population is sampled so that it covers a good representation of the updating solution space. Within the context of this chapter, the size of the population should depend on the nature of the problem, which is determined by the number of variables in the finite-element model that are deemed to be in doubt and therefore need to be updated. For example, if there are two variables to be updated, the size of the population must be greater than when there is only one variable to be updated.

3.3.2 Crossover

The crossover operator mixes genetic information in the population by cutting pairs of chromosomes at random points along their length and exchanging the cut sections over. This has a potential for joining successful operators together. Crossover, in this chapter, is an algorithmic operator used to alter the programming of a potential solution to the updating problem from one generation to the other (Gwiazda, 2006).

Crossover occurs with a certain probability. In many natural systems, the probability of crossover occurring is higher than the probability of mutation occurring. One example is a simple crossover technique (Banzhaf *et al.*, 1998; Goldberg, 1989).

For simple crossover, one crossover point is selected, a binary string from the beginning of a chromosome to the crossover point is copied from one parent, and the rest is copied from the second parent. For example, if two chromosomes in binary space a=11001011 and b=11011111 undergo a one-point crossover at the midpoint, then the resulting offspring is c=11001111. For arithmetic crossover, a mathematical operator is performed to make an offspring. For example, an AND operator can be performed on a=11001011 and b=11011111 to form an offspring 11001011. In this chapter, arithmetic crossover is used.

3.3.3 Mutation

The mutation operator picks a binary digit of the chromosomes at random and inverts it. This has a potential of introducing new information to the population, and thereby prevents the genetic-algorithm simulation from being stuck in a local optimum solution. Mutation occurs with a certain probability. In many natural systems, the probability of mutation is low (*i.e.*, less than 1%). In this chapter, binary mutation is used (Goldberg, 1989). When binary mutation is used, a number written in binary form is chosen and one bit value is inverted. For example: the chromosome 11001011 may become the chromosome 11000011.

In this chapter, nonuniform mutation is used. Nonuniform mutation operates by increasing the probability of mutation in such a way that it will be close to 0 as the generation number increases sufficiently. It prevents the population from stagnating in the initial stages of the evolution process, and then permits the algorithm to refine the solution in the end stages of the evolution.

3.3.4 Selection

For every generation, a selection of the proportion of the existing population is chosen to breed a new population. This selection is conducted using the fitness-based process, where solutions that are fitter, as measured by Equation 3.4, are given a higher probability of being selected. Some selection methods rank the fitness of each solution and choose the best solutions, while other procedures rank a randomly chosen sample of the population for computational efficiency.

Many selection functions tend to be stochastic in nature and thus are designed in such a way that a selection process is conducted on a small proportion of less fit solutions. This ensures that diversity of the population of possible solutions is maintained at a high level and, therefore, avoiding convergence on poor and incorrect solutions. There are many selection methods and these include roulette-wheel selection (Mohamed *et al.*, 2008), which is used in this chapter.

Roulette-wheel selection is a genetic operator used for selecting potentially useful solutions in a genetic-algorithm optimization process. In this method, each possible procedure is assigned the fitness function that is used to map the probability of selection with each individual solution. Suppose the fitness f_i is of individual i in the population, then the probability that this individual is selected is:

$$p_i = f_i \bigg/ \sum_{j=1}^{N} f_j \tag{3.6}$$

Here, N is the total population size.

This process ensures that candidate solutions with a higher fitness have a lower probability so that they may eliminate those with a lower fitness. By the same token, solutions with low fitness have a low probability of surviving the selection process. The advantage of this is that even though a solution may have low fitness, it may still contain some components that may be useful in the future.

The processes described result in the subsequent generation of a population of solutions that is different from the previous generation and that has an average fitness that is higher than the previous generation.

3.3.5 Termination

The process described is repeated until a termination condition has been achieved, either because a desired solution that satisfies the objective function in Equation 3.4 was found or a specified number of generations has been reached or the solution's fitness converged (or any combination of these).

The process described above can be written in pseudo-code, as shown in Algorithm 3.1 (Goldberg, 1989). Table 3.1 shows the operations, types and parameters in the implementation of a genetic algorithm. This table indicates that in a genetic-algorithm implementation, there are many choices that ought to be made.

For example, for genetic-algorithm representation, a choice has to be made between a binary and a floating-point representation. In this chapter, binary representation is used. Given this choice, a bit size must be chosen. For this chapter, a 16-bit binary representation was chosen.

Algorithm 3.1 An algorithm for implementing a genetic algorithm

1. Select the initial population
2. Calculate the fitness of each chromosome in the population using Equation 3.4
3. Repeat
   a. Choose chromosomes with higher fitness to reproduce
   b. Generate a new population using crossover and mutation to produce offspring
   c. Calculate the fitness of each offspring
   d. Replace the low fitness section of the population with offspring
4. Repeat until termination

For the initialization process, a choice has to be made for the population size. Table 3.1 illustrates that in the implementation of a genetic algorithms, the difficulty is that there are many choices to be made and there is no direct methodology on how these choices must be made, so these choices tend to be arbitrary.

3.4 Nelder–Mead Simplex Optimization Method

The Nelder–Mead simplex method was developed by Nelder and Mead in 1965. It is a method for minimizing an objective function, Equation 3.4 in this chapter, in a multi-dimensional space (Nelder and Mead, 1965). The Nelder–Mead simplex method produces a new test position by extrapolating the behavior of the objective function measured at each test position arranged as a simplex. An N-simplex is an n-dimensional structure that is analogous to a triangle. The simplest step is intended to substitute the worst position with a position reflected through the centroid of the remaining N points. If this position is better than the best current point, then it can be used to attempt to stretch it exponentially out along this line. Alternatively, if this current position is not sufficiently improved over the prior value, then the process is used in stepping across a valley and the simplex is shrunk towards an improved position. The Nelder–Mead simplex method can converge to a nonstationary position. The algorithm then chooses to substitute one of these test positions with new test positions and the technique progresses until convergence. This technique has been used successfully and Schlune *et al.*, (2009) used the Nelder–Mead simplex method for finite-element-model updating in bridges while Zapico *et al.* (2003) applied the Nelder–Mead simplex method for finite-element-model updating in a small-scale bridge.

Table 3.1. Operations, types and parameters for the implementation of a genetic algorithm

Operation	Types	Parameters to select
Genetic representation	Binary, floating point	Bit size
Initialization	Population, random seed	Population size, Distribution of the random seed
Crossover	Arithmetic, simple, one-point, two-point, uniform	Probability of crossover
Mutation	Nonuniform, binary	Probability of mutation
Fitness function evaluation	Problem specific	
Selection	Roulette wheel, tournament selection	
Reproduction	Two parents, three parents reproduction	

Decouvreur *et al.* (2004) updated two-dimensional acoustic models with the constitutive relation error method using a hybrid simplex artificial bee colony algorithm combining the Nelder–Mead simplex method with the artificial bee-colony algorithm. Begambre and Laier (2009) used a hybrid particle-swarm optimization and the Nelder–Mead simplex method in structural-damage identification.

3.5 Example 1: Simple Beam

The aluminum beam shown in Chapter 2 was used to test Nelder–Mead simplex and a genetic-algorithm optimization methods for finite-element-model updating. This beam had the following dimensions: length: 1.1 m, width: 29.2 mm and thickness: 9.6 mm. This beam had holes of diameter 5.8 mm located at the centers of elements 2 to 9 and, therefore, was difficult to model. Further details of this beam are reported in Marwala (1997). The beam was tested, freely suspended using elastic rubber bands. The beam was excited using an electromagnetic shaker and the response was measured using an accelerometer. The beam was also modeled using the structural dynamics toolbox (Balmès, 1997) and it was divided into 11 elements. The finite-element model used Euler–Bernoulli beam elements. It was excited at various positions and acceleration was measured at 10 different positions.

A set of 10 frequency-response functions were calculated and a roving accelerometer was used for the testing. The moduli of elasticity of these elements were used as updating parameters. When the finite-element-model updating was implemented the moduli of elasticity was restricted to vary from 6×10^{10} to 8×10^{10} N m^{-2}. The weighting factors, in the first terms of Equation 3.4, were calculated for each mode as the square of the error between the measured natural frequency and the natural frequency calculated from the initial model. The weighting function for the second term in Equation 3.4 was set to 0.75. When the

Nelder–Mead simplex and genetic-algorithm techniques were implemented for finite-element-model updating, the results shown in Table 3.2 were obtained. The results for the NM are slightly different from those in Chapter 2 because of the different objective functions used.

Table 3.2 Results showing measured frequencies of the beam, the initial frequencies and the frequencies obtained when the finite-element model of a beam was updated using NM and GA

Modes	Measured frequency (Hz)	Initial frequency (Hz)	Frequencies from the NM updated model (Hz)	Frequencies from GA the updated model (Hz)
1	41.5	42.3	39.7	40.4
2	114.5	117.0	114.7	114.2
3	224.5	227.3	221.3	223.3
4	371.6	376.9	370.3	371.5

The genetic algorithm was run for a population of 50 and over 200 generations. Arithmetic crossover with a probability of 40% and a nonuniform mutation with a probability of 0.5% were also implemented. These choices were arrived at after numerous experiments.

When a modulus of elasticity of 7×10^{10} N m^{-2} was assumed, the error between the first measured natural frequency and that from the initial finite-element model was 1.9%. When the Nelder–Mead simplex method was used for finite-element-model updating, this error was increased to 4.3% but when using the genetic-algorithm approach it was increased to 2.7%.

The error between the second measured natural frequency and that from the initial model was 2.2%. When the Nelder–Mead simplex method was used, this error was reduced to 0.2% and using the genetic algorithm it was reduced to 0.3%.

The error between the measured data and the initial finite-element model in the third natural frequencies was 1.2%. When the Nelder–Mead simplex method was used, this error was increased to 1.4% and using the genetic algorithm it was reduced to 0.5%.

The error between the fourth measured natural frequency and that from the initial model was 1.4%. When the Nelder–Mead simplex method was used, this error was reduced to 0.3% and using the genetic algorithm it was reduced to 0.0%.

Overall, the genetic algorithm gave the best results with an average error of 0.9%, while the Nelder–Mead simplex method gave an average error of 1.6%. On average, both methods improved when compared to the average error between the initial finite-element model and the measured data.

The updated models implemented were also validated on the mode shapes they predicted. To make this assessment possible, the MAC was used and the results are shown in Table 3.3. The mean of the diagonal of the MAC vector was used to compare the mode shapes predicted by the updated and initial finite-element models to the measured mode shapes. The average value of 1.0 indicates that the mode shapes are properly correlated. The average MAC calculated between the mode shapes from an initial finite-element model and the measured mode shapes was 0.9986. When the average MAC was calculated between the measured data and data obtained from the updated finite-element models, it was observed that the genetic-algorithm updated finite-element models give improved averages for the diagonals of the MAC matrices of 0.9989, while the Nelder–Mead simplex method gave an average of 0.9988.

Table 3.3 Beam results showing the MAC calculated between measured mode shapes and the initial finite-element model, the NM updated finite-element model and the GA updated finite-element model

Method	MAC
Initial model	0.9986
NM	0.9988
GA	0.9989

3.6 Example 2: Unsymmetrical H-shaped Structure

The unsymmetrical H-shaped aluminum structure shown in Chapter 2 was also used to validate the proposed method. This structure was previously used by Marwala and Heyns (1998) and by Marwala (1997). This structure has three thin cuts of 1 mm that go half-way through the cross-section of the beam. These cuts were introduced to elements 3, 4 and 5. The structure with these cuts was used so that the initial finite-element model gave data that were far from the measured data and, thereby test the proposed procedure on a difficult finite-element-model updating problem. The structure was suspended using elastic rubber bands. The structure was excited using an electromagnetic shaker and the response was measured using an accelerometer. The structure was divided into 12 elements.

It was excited with acceleration measured at 15 positions. The structure was tested freely suspended, and a set of 15 frequency-response functions were calculated. A roving accelerometer was used for the testing. The mass of the accelerometer was found to be negligible compared to the mass of the structure. As in the previous example, the finite-element model was constructed using the structural dynamics toolbox (Balmès, 1997) using the Euler–Bernoulli beam elements. The finite-element model contained 12 elements. As in the previous example, the moduli of elasticity of these elements were used as updating parameters that were restricted to fall in the interval from 6×10^{10} to 8×10^{10} N m^{-2}. The Nelder–Mead simplex method and a genetic algorithm were implemented as in the previous example.

The results obtained when the Nelder–Mead simplex method and genetic algorithms were used for finite-element-model updating, are shown in Table 3.4. Table 3.4 shows the measured natural frequencies, initial natural frequencies and natural frequencies obtained by the Nelder–Mead simplex method and the genetic-algorithm updated finite-element models. The results of the NM found in this chapter are different from those in Chapter 2 because of the different updating objective function used.

Table 3.4 Results from an unsymmetrical H-shaped structure showing measured frequencies, the initial frequencies and the frequencies obtained when the finite-element model was updated using the NM and the GA

Modes	Measured frequency (Hz)	Initial frequency (Hz)	Frequencies from the NM updated model (Hz)	Frequencies from the GA updated model (Hz)
1	53.9	56.2	52.1	53.9
2	117.3	127.1	119.4	120.1
3	208.4	228.4	212.4	211.3
4	254.0	263.4	251.3	253.4
5	445.1	452.4	433.6	438.6

The error between the first measured natural frequency and that from the initial finite-element model, obtained when a modulus of elasticity of 7×10^{10} N m^{-2} was assumed, was 4.3%. When the Nelder–Mead simplex method was used for the finite-element-model updating this error was reduced to 3.3%, and the genetic-algorithm approach reduced this error to 0%.

The error between the second measured natural frequency and that from the initial model was 8.4%. When the Nelder–Mead simplex method was used, the error was reduced to 1.8% and the genetic algorithm reduced this error to 2.4%. The error in the third natural frequencies between the measured data and the initial finite-element model was 9.6%. When the Nelder–Mead simplex method was used, this error was reduced to 1.9% and using a genetic algorithm it was reduced to 1.4%.

The error between the fourth measured natural frequency and that from the initial model was 3.7%. When the Nelder–Mead simplex method was used this error was reduced to 1.0%, and using the genetic algorithm it was reduced to 0.2%. The error between the fifth measured natural frequency and that from the initial model was 1.6%.

When the Nelder–Mead simplex method was used this error was increased to 4.1% and the genetic algorithm, increased it to 1.4%. Overall, the genetic algorithm gave the best results with an average error calculated over all the five natural frequencies of 1.3%, while the Nelder–Mead simplex method gave an average error of 3.0%. On average, both methods improved when compared with the average error between the initial finite-element model and the natural frequencies.

As in the previous example, the updated models were validated on the mode shapes they predicted using the MAC. The results appear in Table 3.5. Table 3.5 shows that the NM- and GA-updated finite-element models gave improved averages for the diagonals of the MAC matrices. Therefore, the GA gave the best MAC.

Table 3.5 Results of the unsymmetrical H-shaped structure showing MAC calculated between measured mode shapes and the initial finite-element model, the NM-updated finite-element model and the GA-updated finite-element model

Method	MAC
Initial model	0.8394
NM	0.8404
GA	0.8419

3.7 Conclusion

In this study, the Nelder–Mead simplex method and a genetic algorithm were implemented for finite-element-model updating. When these techniques were tested on a simple beam and an unsymmetrical H-shaped structure, it was observed on average that the genetic algorithm gave more accurate results than the Nelder–Mead simplex method. The Nelder–Mead simplex method was found to be more computationally efficient than the a genetic algorithm.

3.8 Future Work

This chapter introduced a genetic algorithms for finite-element-model updating. For further work, methods should be identified for designing the entire finite-element updating algorithm system for maximum efficiency in terms of accuracy of measured data estimation. Another issue that needs to be addressed is how to determine theoretically the relationship between the number of updating parameters and the accuracy of the finite-element model, given the level of errors in the measured data. The conclusions reached in this chapter are highly dependent on the nature of the data used in the analysis. Therefore, further statistical tests need to be conducted to ensure that the conclusions reached are not dependent on the data used. In particular, a relationship between the nature of data and the finite-element-model-updating process to be used must be identified.

References

Abdella M, Marwala T (2006) The Use of A genetic algorithms and Neural Networks to Approximate Missing Data in Database. Comput and Informatics 24:1001–1013

Akula VR, Ganguli R (2003) Finite Element Model Updating for Helicopter Rotor Blade Using A genetic algorithm. AIAA J. doi: 10.2514/2.1983

Allemang RJ, Brown DL (1982) A Correlation Coefficient for Modal Vector Analysis. In: Proc of the 1st Int Modal Anal Conf:01–18

Almeida FS, Awruch AM (2009) Design Optimization of Composite Laminated Structures Using A genetic algorithms and Finite Element Analysis. Compos Struct 88:443–454

Arifovic J, Gençay R (2001) Using A genetic algorithms to Select Architecture of a Feedforward Artificial Neural Network. Physica A: Statistical Mech and its App 289:574–594

Balamurugan R, Ramakrishnan CV, Singh N (2008) Performance Evaluation of a Two Stage Adaptive A genetic algorithm (TSAGA) in Structural Topology Optimization. Appl Soft Comput 8:1607–1624

Balmès E (1997) Structural Dynamics Toolbox User's Manual Version 2.1 Sèvres, France: Scientific Software Group

Banzhaf W, Nordin P, Keller R, Francone, F (1998) Genetic Programming-An Introduction: On the Automatic Evolution of Computer Programs and its Applications. Morgan Kaufmann Publishers, California

Begambre O, Laier JE (2009) A Hybrid Particle Swarm Optimization – Simplex Algorithm (PSOS) for Structural Damage Identification. Adv in Eng Softw 40:883–891

Canyurt OE, Kim HR, Lee KY (2008) Estimation of Laser Hybrid Welded Joint Strength by Using A genetic algorithm Approach. Mech of Mater 40: 825–831

Crossingham B, Marwala T (2007) Using A genetic algorithms to Optimise Rough Set Partition Sizes for HIV Data Analysis. Stud in Comput Intell 78:245–250

Decouvreur V, Bouillard PH, Deraemaeker A, Ladevèze P (2004) Updating 2D Acoustic Models with the Constitutive Relation Error Method. J of Sound and Vib 278:773–787

Doebling SW, Farrar CR, Prime MB, Shevitz DW (1996) Damage Identification and Health Monitoring of Structural and Mechanical Systems from Changes in their Vibration Characteristics: A Literature Review. Los Alamos National Laboratory Report LA-13070-MS

Ewins DJ (1995) Modal Testing: Theory and Practice. Research Studies Press, Letchworth

Forrest S (1996) A genetic algorithms. ACM Comput Surv 28:77–80

Fourie PC, Groenwold AA (2000) Particle Swarms in Size and Shape Optimization. In: Proc of the Int Workshop on Multidiscip Des Optim: 97–106

Fourie PC, Groenwold AA (2001) Particle Swarms in Topology Optimization. In: Ext Abstr of the 4th World Congr of Struct and Multidiscip Optim:52–53

Franulović M, Basan R, Prebil I (2009) A genetic algorithm in Material Model Parameters' Identification for Low-cycle Fatigue. Comput Mater Sci 45: 505–510

Friswell MI, Mottershead JE (1995) Finite Element Model Updating in Structural Dynamics. Kluwer Academic Publishers Group, Norwell

Goldberg DE (1989) A genetic algorithms in Search, Optimization and Machine Learning. Addison-Wesley, Reading

Gwiazda, TD (2006) A genetic algorithms Reference Vol.1 Cross-over for Single-objective Numerical Optimization Problems. Adobe eBook, Lomianki

Houck CR, Joines JA, Kay MG (1995) A genetic algorithm for Function Optimisation: A MATLAB implementation. Tech. Rep. NCSU-IE TR 95-09, North Carolina State University

Hulley G, Marwala T (2007) A genetic algorithm Based Incremental Learning for Optimal Weight and Classifier Selection. Comp Models for Life Sci. Am Inst of Phys Ser 952:258–267

Kubalík J, Lazanský J (1999) A genetic algorithms and their Testing. AIP Conf Proc 465:217–229

Kwak HG, Kim J (2009) An Integrated A genetic algorithm Complemented with Direct Search for Optimum Design of RC Frames. Computer-Aided Des 41:490–500

Levin RI, Lieven NAJ (1998) Dynamic Finite Element Model Updating using Simulated Annealing and A genetic algorithms. Mech Syst and Signal Process 12:91–120

Marwala T (1997) A Multiple Criterion Updating Method for Damage Detection on Structures. Masters Thesis, University of Pretoria

Marwala T (2002) Finite Element Updating Using Wavelet Data and A genetic algorithm. AIAA J of Aircr 39:709–711

Marwala T (2003) Control of Fermentation Process Using Bayesian Neural Networks and A genetic algorithm. In: Proc of the African Control Conf: 449–454

Marwala T (2004) Finite Element Model Updating Using Response Surface Method. 45th AIAA/ASME/ASCE/AHS/ASC Structures, Structural Dyn & Mater Conf: Paper AIAA-2004-2005

Marwala T (2005) Finite Element Model Updating Using Particle Swarm Optimization. Int J of Eng Simul 6:25–30

Marwala T (2007) Bayesian Training of Neural Network Using Genetic Programming. Pattern Recognit Lett 28:1452–1458

Marwala T, Chakraverty S (2006) Fault Classification in Structures with Incomplete Measured Data Using Autoassociative Neural Networks and A genetic algorithm. Curr Sci 90:542–548

Marwala T, de Wilde P, Correia L, Mariano P, Ribeiro R, Abramov V, Szirbik N, Goossenaerts J (2001) Scalability and Optimisation of a Committee of Agents Using A genetic algorithm. In: Proc of the Int Symp on Soft Comput and Intell Syst for Ind

Marwala T, Heyns PS (1998) A Multiple Criterion Method for Detecting Damage on Structures, AIAA J 195:1494–1501

Michalewicz, Z (1996) A genetic algorithms + Data Structures = Evolution Programs. Springer-Verlag, New York

Mitchell M (1996) An Introduction to A genetic algorithms. MIT Press, Cambridge

Mohamed AK, Nelwamondo FV and Marwala T. (2008) Estimation of Missing Data: Neural Networks, Principal Component Analysis and A genetic algorithms. In: Proc of the 12th World Multi-Conf on Syst, Cybern and Inform:36–41

Nelder JA, Mead R (1965) A Simplex Method for Function Minimization. Comput J 7:308–313

Oh S, Pedrycz W (2006) Genetic Optimization-driven Multi-layer Hybrid Fuzzy Neural Networks. Simul Model Pract and Theory 14:597–613

Paluch B, Grédiac M, Faye A (2008) Combining a Finite Element Programme and a genetic algorithm to Optimize Composite Structures with Variable Thickness. Compos Struct 83:284–294

Park BJ, Choi HR, Kim HS (2003) A Hybrid A genetic algorithm for the Job Shop Scheduling Problems. Comput and Ind Eng 45:597–613

Pendharkar PC, Rodger JA (1999) An Empirical Study of Non-binary A genetic algorithm-based Neural Approaches for Classification. In: Proc of the 20th Int Conf on Inf Syst:155–165

Perera R, Ruiz A (2008) A Multistage FE Updating Procedure for Damage Identification in Large-scale Structures Based on Multi-objective Evolutionary Optimization. Mech Syst and Signal Process 22:970–991

Perera R, Ruiz A, Manzano C (2009) Performance Assessment of Multi-criteria Damage Identification A genetic algorithms. Comput & Struct 87:120–127

Roy T, Chakraborty D (2009) Optimal Vibration Control of Smart Fiber Reinforced Composite Shell Structures Using Improved A genetic algorithm. J of Sound and Vib 319:15–40

Schlune H, Plos M, Gylltoft K (2009) Improved Bridge Evaluation Through Finite Element Model Updating Using Static and Dynamic Measurements. Eng Struct 31:1477–1485

Stern H, Chassidim Y, Zofi M (2006) Multi-agent Visual Area Coverage Using a New A genetic algorithm Selection Scheme. Eur J of Oper Res 175:1890–1907

Tettey T, Marwala T (2006) Controlling Interstate Conflict Using Neuro-fuzzy Modeling and A genetic algorithms. In: Proc of the 10th IEEE Int Conf on Intell Eng Syst:30–44

Tibshirani R (1996) Regression Shrinkage and Selection via the Lasso (PostScript). J of the R Statistical So, Ser B (Methodology) 58:267–288

Tu Z, Lu Y (2008) Finite Element Model Updating Using Artificial Boundary Conditions with A genetic algorithms. Comput & Struct 86:714–727

Vose, MD (1999) The Simple A genetic algorithm: Foundations and Theory, MIT Press, Cambridge

Zapico JL, González MP, Friswell MI, Taylor CA, Crewe AJ (2003) Finite Element Model Updating of a Small Scale Bridge. J of Sound and Vib 268:993–1012

Zhang H, Ishikawa M (2004) A Solution to Combinatorial Optimization with Time-varying Parameters by a Hybrid A genetic algorithm. Int Congr Ser 1269:149–152

Chapter 4

Finite-element-model Updating Using Particle-swarm Optimization

Abstract. This chapter implements the particle-swarm-optimization method for finite-element-model updating. This method is tested on a simple beam and an unsymmetrical H-shaped structure and compared with a method that is based on a genetic-algorithm optimization technique. It is observed that, on average, the particle-swarm-optimization method gives a more accurately updated finite-element model than does the genetic-algorithm method.

Keywords: particle-swarm optimization, genetic algorithm, objective function

4.1 Introduction

In Chapter 3, the genetic algorithm (GA) method was implemented for finite-element-model updating. This method was tested on a simple beam and an unsymmetrical H-shaped structure and compared to a method that is based on the Nelder–Mead (NM) simplex method. It was noted that, on average, the genetic-algorithm method gives a more accurately updated finite-element-updated model than does the Nelder–Mead simplex method. This is because the GA, which is a population-based method, can escape from a local optimum solution better than the Nelder–Mead simplex method. This is because the genetic algorithm contains a mutation step whose primary function is to give the genetic algorithm the ability to escape from a local optimum solution.

As described in previous chapters, finite-element-model updating is a process of tuning a finite-element model so that it can better reflect the observed data from the physical structure being modeled (Friswell and Mottershead, 1995). The finite-element-model-updating process can be viewed as a process of identifying the highly uncertain parameters of the finite-element model from the measured data or, alternatively, as a process of enabling the finite-element model to better predict the measured data. As indicated before, this process is fundamentally an optimization approach where the design variables are the parameters of the finite-element model that are deemed to be in doubt and the objective function is an equation that characterizes some distance between the finite-element-model predictions and the measured data.

As mentioned in previous chapters, many optimization methods have been used for finite-element-model updating. Levin and Lieven (1998) introduced genetic-algorithm and simulated-annealing methods for finite-element-model updating. Tu and Lu (2008) successfully implemented finite-element-model updating using artificial boundary conditions using a genetic algorithm, while Schlune *et al.* (2009) improved the finite-element model by using static and dynamic measurements. The main drawback with a genetic-algorithm method is the complexity of its implementation, which entails the choice of:

- genetic representation using floating point or binary;
- the population size; and
- the type as well as the probability of crossover, mutation and selection.

Because of these limitations, this chapter presents a particle-swarm-optimization method for finite-element-model updating, which is a population-based approach that was inspired by a mathematical description of the swarming of birds. In Chapter 3, the genetic algorithm was found to be better than the Nelder–Mead simplex method, so this chapter will compare particle-swarm optimization with a genetic-algorithm technique. The advantages of particle-swarm-optimization method are (Perez and Behdinan, 2007):

- it handles the scaling of the design variables (*i.e.*, finite-element-updating parameters) well;
- it is easy to execute;
- it can be implemented easily using parallel processing;
- it does not require the calculation of the objective function derivatives;
- it contains a small number of adjustable parameters; and
- it is an effective algorithm for identifying a global optimum solution.

The disadvantage of particle-swarm optimization is that it is not good at searching for a local solution and therefore it may converge prematurely.

As described in previous chapters, there are two approaches used in finite-element-model updating: direct methods and iterative methods (Friswell and Mottershead, 1995). Direct methods that use the modal properties, are computationally efficient to implement and do reproduce the measured modal data exactly. Furthermore, they do not take into account the updated physical parameters. Iterative procedures use changes in physical parameters to update finite-element models and produce models that are physically realistic.

Iterative methods that use modal properties for finite-element-model updating will be implemented in this chapter. The approach adopted in this regard avoids the expansion or the reduction of the mass and stiffness matrices of the finite-element model. The finite-element models are updated so that the measured modal properties match the finite-element model's predicted modal properties using the particle-swarm-optimization method and the genetic algorithm and the two methods are compared using a simple beam and an unsymmetrical H-shaped structure as applications.

4.2 Mathematical Background

In this chapter, as in the previous chapter, modal properties, *i.e.*, natural frequencies and mode shapes are used as a basis for finite-element-model updating (Ewins, 1995). Modal properties are related to the physical properties of the structure. All elastic structures may be described in terms of the relationship between their distributed mass, damping and stiffness matrices as well as force using Newton's law of motion. The finite-element model gives a representation of this relationship. From this finite-element model, the relationship between the changes in the mass and stiffness matrices as well as the modal properties of the structure can be quantified. The correct mass and stiffness matrices of a system can therefore be identified from the measured data. Data are usually measured in the time domain. A process called *modal analysis* is used to extract modal properties that are the natural frequencies and mode shapes explained in Chapter 1. This extraction process uses the following equation (Ewins, 1995):

$$H_{kl}(\omega) = \sum_{i=1}^{N} \frac{-\omega^2 \phi_k^i \phi_l^i}{-\omega^2 + 2\zeta_i \omega_i \omega j + \omega_i^2} \tag{4.1}$$

Here, $H_{kl}(\omega)$ is the frequency-response function (FRF) due to excitation at k and response measurement at l; ω is the frequency point; ω_i is the ith natural frequency point; N is the number of modes; and ζ_i is the damping ratio of mode i.

In Equation 4.1 the FRF, which is known, is defined as the ratio of the Fourier-transformed response to the Fourier-transformed force (Ewins, 1995). Then, the natural frequencies, mode shapes and damping ratios are deduced by minimizing the difference between the frequency-response function as calculated in Equation 4.1 and the measured FRF.

As in Chapter 3, the finite-element-model updating process is achieved by identifying the correct mass and stiffness matrices that ensure that the finite-element model predicts the measured data to a sufficient accuracy. In the light of the measured data, the correct mass and stiffness matrices can be obtained by identifying the correct moduli of elasticity of various sections of the structure under consideration. In this chapter, to correctly identify the moduli of elasticity that would give the updated finite-element model, the following objective function, which measures the distance between measured modal data and finite-element-model-calculated modal data, is minimized:

$$E = \sum_{i=1}^{N} \gamma_i \left(\frac{\omega_i^m - \omega_i^{calc}}{\omega_i^m} \right)^2 + \beta \sum_{i=1}^{N} \left(1 - diag(MAC(\phi_i^{calc}, \phi_i^m))\right) \tag{4.2}$$

Here, M indicates a measured quantity; calc indicates a calculated quantity; ω_i is the ith natural frequency; ϕ_i is the ith mode shape vector; N is the number of modes; γ_i is the weighting factor that measures the relative distance between the initial estimated natural frequencies for mode i and the target frequency of the same mode; β is the weighting function on the mode shapes; MAC is the modal assurance criterion (Allemang and Brown, 1982); and diag(MAC)$_i$ is the ith diagonal element of the MAC matrix.

Conventionally, Equation 4.2 includes the regularization parameters (Hua *et al.*, 2009; Weber *et al.*, 2009), which is a process of introducing an additional penalty function to solve an ill-posed problem or prevent over-fitting the model. This ensures that the finite-element model is not too tuned to reflect the measured data at the expense of the physics in the finite-element model. In this chapter, to deal with this potential problem, particular attention is paid to the choice of the bounds of the parameters to be updated with an emphasis on the engineering judgment of the user.

In Equation 4.2, the first term has the function of ensuring that the natural frequencies predicted by the finite-element model are as close to the measured ones as possible, while the second-term ensures that the mode shapes between measurements and finite-element-model prediction are correlated. The MAC is a measure of the correlation between two sets of mode shapes of the same dimension and can be represented mathematically as follows (Allemang and Brown, 1982):

$$MAC_{cdr} = \frac{\left|\{\phi_{cr}\}\{\phi_{dr}^{*}\}\right|^{2}}{\{\phi_{cr}\}^{T}\{\phi_{cr}^{*}\}\{\phi_{dr}\}^{T}\{\phi_{dr}^{*}\}} \tag{4.3}$$

Here, MAC is the modal assurance criterion; c is for reference; d is the degree of freedom; r is the mode; T is the transpose; * is the complex conjugate; and { } is a vector.

In this chapter the process outlined in Figure 4.1 is followed. In this figure, the physical structure is modeled using the finite-element model. In parallel with this, measurements are taken from the structure and the distance between the measured results and those from the finite-element model is calculated. If the distance between these two sets of results is sufficiently small, then there is no need for finite-element-model updating. If this is not the case, then the finite-element model is parameterized by identifying parameters that are deemed uncertain. After this process, the objective function, which is represented in Equation 4.2, is constructed. This objective function is solved in this chapter using a genetic algorithm and particle-swarm optimization. The theory of particle-swarm optimization is discussed in the next section.

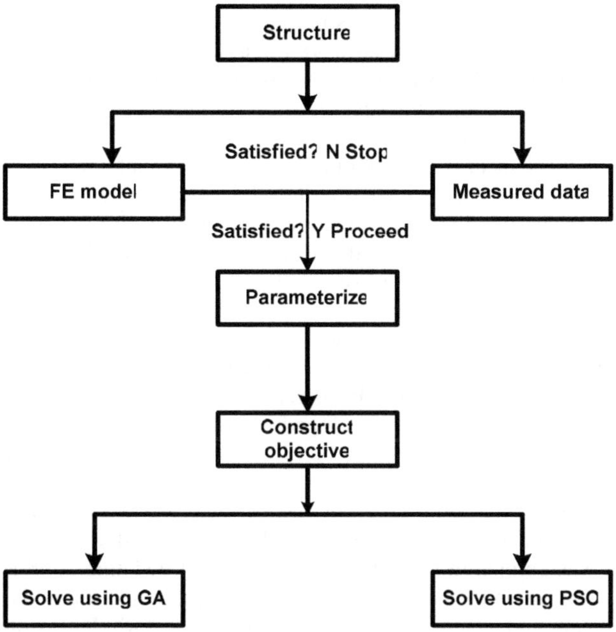

Figure 4.1 Outline of the procedures followed in this chapter

4.3 Particle-swarm Optimization

This chapter uses the particle-swarm optimization (PSO) method to solve Equation 4.2. PSO is a stochastic, population-based evolutionary algorithm that is widely used for optimization. It is based on socio-psychological principles that are inspired by swarm intelligence, which offers understanding into social behavior and has contributed to engineering applications. Society enables an individual to maintain cognitive robustness through influence and learning and individuals learn to tackle problems by communicating and interacting with other individuals and, thereby, develop a generally similar way of tackling problems (Engelbrecht, 2005). Thus, swarm intelligence is driven by two factors:

1. group knowledge; and
2. individual knowledge.

Each member of a swarm always acts by balancing between its individual knowledge and the group knowledge.

To solve optimization problems using particle-swarm optimization, a fitness function is constructed to describe a measure of the desired outcome. In this chapter the fitness function is the distance between the data predicted from the finite-element model and the measured data with the uncertain parameters as the design variables. To reach an optimum updated finite-element, a social network representing a population of possible solutions is defined and randomly generated. The individuals within this social network are assigned neighbors to interact with. These individuals are called particles, hence the name particle-swarm optimization. Thereafter, a process to update these particles is initiated. This is conducted by evaluating the fitness of each particle. Each particle can remember the location where it had its best success as measured by the fitness function. The best solution of the particle is named the *local best* and each particle makes this information on the local best accessible to their neighbors and in turn observe their neighbors' success. The process of moving in the search space is guided by these successes and the population ultimately converges by the end of the simulation on an optimum solution.

The particle-swarm optimization technique was developed by Kennedy and Eberhart (1995). This procedure was inspired by algorithms that model the "flocking behavior" seen in birds. Researchers in artificial life (Reynolds, 1987; Heppner and Grenander, 1990) developed simulations of bird flocking. In the context of optimization, the concept of birds finding a roost is analogous to a process of finding an optimal solution. Particle-swarm optimization has been very successful in optimizing complex problems. Marwala (2005) used particle-swarm optimization to improve finite-element models to better reflect the measured data. This method was compared to a finite-element-model updating approach that uses simulated annealing and a genetic algorithm. The presented methods were tested on a simple beam and an unsymmetrical H-shaped structure. It was observed that, on average, the particle-swarm-optimization method gave the most accurate results followed by simulated annealing and then the genetic algorithm.

Dindar (2004) as well as Dindar and Marwala (2004) successfully used particle-swarm optimization to optimize the structure of the committee of neural networks. The results obtained from the optimized networks were found to be better than both unoptimized networks and the committee of networks. Ransome *et al.* (2005) as well as Ransome (2006) successfully used particle-swarm optimization to optimize the position of a patient during radiation therapy. In this application, a patient-positioning system integrating a robotic arm was designed for proton-beam therapy. A treatment image was aligned with a predefined reference image and this was attained by aligning the radiation and reference field boundaries and then registering the patient's anatomy relative to the boundary. Methods for both field boundary and anatomy alignment, including particle-swarm optimization, were implemented. It was found that the particle-swarm optimization was successful in overcoming problems in existing solutions.

Arumugam and Rao (2008) successfully implemented multiple-objective particle-swarm optimization for molecular docking. Multi-objective particle-swarm optimization is a technique that aims at solving more than one objective function. For example, in this chapter as well as in the previous chapter, the finite-element updating process is intended to solve a single objective function, represented by Equation 4.2.

Arya *et al.* (2007) successfully used particle-swarm optimization and singular-value decomposition to design neuro-fuzzy networks. This process was aimed at identifying the optimal neuro-fuzzy networks whose objective is to accurately model the data. Berlinet and Roland (2008) introduced a particle-swarm optimization algorithm for permutation flow-shop scheduling problems. Scheduling problems have design variables that are integers and, if there are many of these variables, the combinatorial nature of these problems makes the process of finding solutions to these problems extremely difficult.

Jarboui *et al.* (2008) introduced and successfully applied particle-swarm optimization where a distribution vector was used in the update of the velocities of particles. Other successful applications of particle-swarm optimization include Jiang *et al.* (2007) who successfully applied it to a conceptual design; Kathiravan and Ganguli (2007) in power systems; Lian *et al.* (2008) in training radial basis functions that were trained to identify chaotic systems; Lin *et al.* (2008) in inverse radiation problems; Qi *et al.* (2008) in scheduling problems and Guerra and dos S. Coelho (2008) in modeling of nonlinear filters.

Brits *et al.* (2007) presented a particle-swarm optimization technique aimed at locating and refining multiple solutions to problems with multi-modal characteristics. The presented method extended the unimodal nature of the standard particle-swarm-optimization approach by using many swarms from the initial population. A different solution was represented by a subswarm and was individually optimized. Thereby, each set of particle in the swarm represent a possible solution. When implemented experimentally it was found to successfully locate all optima.

Sha and Hsu (2006) proposed a hybrid particle-swarm optimization for the job-shop problem that operated in discrete space rather than in continuous space. Due to the nature of the discrete space, the particle-swarm-optimization algorithm was improved through the particle-position representation, particle movement, and particle velocity. The particle-position was then represented based on a preference list, particle movement on the swap operator, and particle velocity on the tabu list concept. Giffler and Thompson (1960) implemented a heuristic to decode a particle position into a schedule, while the tabu search improved the solution quality.

Finite-element-model updating is essentially a process of estimating parameters in the finite-element model given the measured data. Zhang *et al.* (2009) used particle-swarm optimization to estimate calibration parameters in a geotechnical model, while Thakker *et al.* (2009) applied particle-swarm optimization for parameter estimation in a numerical model, whereas Liu *et al.* (2008) applied particle-swarm optimization for parameter estimation in a permanent magnet.

Particle-swarm optimization is implemented by finding a balance between searching for a good solution and exploiting other particles' success. If the search for a solution is too limited, the simulation will converge to the first solution encountered, which may be a local optimum position. If the successes of others are not exploited then the simulation will never converge. The particle-swarm optimization approach has the following advantages:

- it is computationally efficient;
- it is simple to implement;
- it has few adjustable parameters when compared to other competing evolutionary programming methods such as a genetic algorithm; and
- it can adapt to explore locally and globally.

When implementing the particle-swarm-optimization method, the simulation is initialized with a population of random candidates, each conceptualized as particles. Each particle is assigned a random velocity and is iteratively moved through the particle space. At each step, the particle is attracted towards a region of the best-fitness function by the location of the best fitness achieved so far in the population.

On implementing the standard particle-swarm optimization, each particle is represented by two vectors:

- $p_i(k)$, which is the position of particle *i* at step *k*; and
- $v_i(k)$, which is the velocity of particle *i* at step *k*.

Initial positions and velocities of particles are randomly generated and the subsequent positions and velocities are calculated using the position of the best solution that a particular particle has encountered during the simulation called $pbest_i$ and the best particle in the swarm, which is called $gbest(k)$. The subsequent velocity of a particle *i* can be identified using the following equation:

$$v_i(k+1) = \gamma v_i(k) + c_1 r_1 (pbest_i - p_i(k)) + c_2 r_2 (gbest(k) - p_i(k)) \quad (4.4)$$

where γ is the inertia of the particle, c_1 and c_2 are the "trust" parameters, and r_1 and r_2 are random numbers between 0 and 1.

In Equation 4.4, the first term is the current motion, the second term is the particle-memory influence and the third term is the swarm influence. The subsequent position of a particle *i* can be calculated using these equations:

$$p_i(k+1) = p_i(k) + v_i(k+1) \quad (4.5)$$

The inertia of the particle controls the impact of the previous velocity of the particle on the current velocity. These parameters control the exploratory properties of the simulation with a high value of inertia encouraging global exploration, while a low value of the inertia encourages local exploration.

The parameters c_1 and c_2 are trust parameters. The trust parameter c_1 indicates how much confidence the current particle has on itself while the trust parameter c_2 indicates how much confidence the current particle has on the successes of the population. The parameters r_1 and r_2 are random numbers between 0 and 1 and they determine the degree to which the simulation should explore the space. In Equation 4.5, it can be seen that the particle-swarm optimization makes use of the velocity to update the position of the swarm.

The position of the particle is updated, based on the social behavior of the particles' population and it adapts to the environment by continually coming back to the most promising region identified. This process is stochastic and it can be summarized as follows:

1. Initialize a population of particles' positions and velocities. The positions of the particles must be randomly distributed in the updating parameter space.
2. Calculate the velocity for each particle in the swarm using Equation 4.4.
3. Update the position of each particle using Equation 4.5.
4. Repeat steps 2 and 3 until convergence.

This process is represented diagrammatically in Figure 4.2.

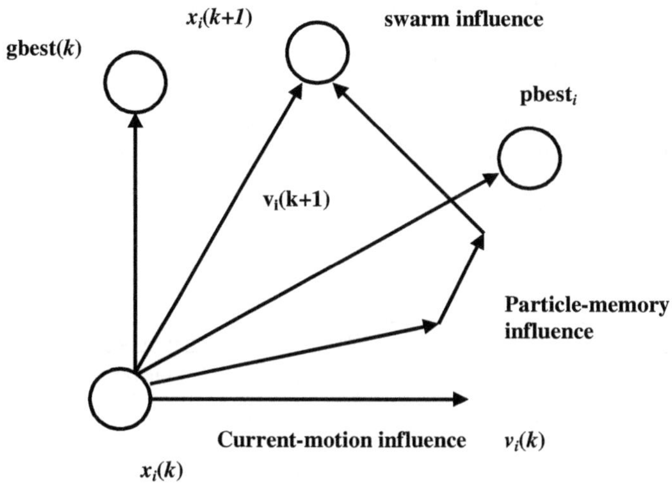

Figure 4.2 Illustration of velocity and particle update in particle-swarm optimization

To improve the performance of particle-swarm optimization as presented above, several additions and modifications of particle-swarm optimization have been presented and implemented. Liu *et al.* (2007) presented a combination of particle-swarm optimization and evolutionary algorithms and applied this method to train recurrent neural networks for time-series prediction.

Janson *et al.* (2007) combined particle-swarm optimization with a gradient-based optimization method and used this to design a composite beam that has the highest possible strength, while Yisu *et al.* (2008) improved particle-swarm optimization by introducing crossover functionality from a genetic algorithm and applied this to controlling hybrid systems.

4.4 Genetic Algorithm (GA)

The finite-element-model-updating method proposed in this chapter also uses a genetic algorithm, as was the case in Chapter 3. Consequently, only a brief overview of genetic algorithms will be presented in this chapter. As explained in Chapter 3, a genetic-algorithm technique is a population-based, probabilistic technique that operates to find a solution to a problem from a population of possible solutions (Kubalik and Lazansky, 1999). It is inspired by Darwin's theory of evolution where members of the population compete to survive and reproduce while the weaker ones die out. Each individual is assigned a fitness value according to how well it meets the objective of solving the problem. New and more evolutionary-fit individual solutions are produced during a cycle of generations, wherein selection and recombination operations, analogous to gene transfer are applied to the current individuals. This continues until a termination condition is met, after which the best individual thus far is considered to be the solution to the problem.

Unlike many optimization algorithms, a genetic algorithm converges to a global optimal solution. In addition, the genetic-algorithm technique has also been proven to be very successful in many applications including finite-element analysis (Marwala, 2003), selecting the optimal neural-network architecture (Arifovic and Gençay, 2001), training hybrid fuzzy neural networks (Oh and Pedrycz, 2006) and solving job-scheduling problems (Park *et al.*, 2003). The differences between particle-swarm optimization and a genetic algorithm are shown in Table 4.1.

Table 4.1 Operations, types and parameters in the implementation of PSO and GA

Operation	Parameters to select PSO	Parameters to select GA
Representation	Floating point	Floating point; bit
Initialization	Population size, distribution of the random seed	Population size, distribution of the random seed
Operations	Velocity and position update, γ, c_1 and c_2, r_1 and r_2	Mutation, crossover and reproduction type and probability

In comparing the particle-swarm optimization to a genetic algorithm Hassan *et al.* (2005) observed that particle-swarm optimization and genetic algorithms have the same effectiveness in finding an optimal solution, and that the particle-swarm optimization-method was computationally better than the genetic algorithm method. The computational efficiency of particle-swarm optimization over the genetic-algorithm method is that there are far too many operations in a genetic algorithm, *e.g.*, crossover, mutation and reproduction, as opposed to velocity and position updating in the particle-swarm-optimization method.

4.5 Example 1: A Simple Beam

The aluminum beam shown in Chapter 2 was used to test the particle-swarm optimization and genetic-algorithm methods for finite-element-model updating. The beam had the following dimensions: length: 1.1 m, width: 29.2 mm and thickness: 9.6 mm. This beam had holes of diameters 5.8 mm located at the centers of elements 2 to 9 and was therefore difficult to model. Further details of this beam are reported in Marwala (1997). The beam was freely suspended using elastic rubber bands. The beam was excited using an electromagnetic shaker and the response was measured using an accelerometer.

The beam was then also modeled using the structural dynamics toolbox (Balmès, 1997) and the beam was divided into 11 elements. The finite-element model used Euler–Bernoulli beam elements. It was excited at various positions and the acceleration was measured at 10 positions. A set of 10 frequency-response functions were calculated and a roving accelerometer was used for testing. The moduli of elasticity of these elements were used in updating the parameters.

When the finite-element-model updating was implemented, the moduli of elasticity were restricted to vary from 6×10^{10} to 8×10^{10} N m^{-2}. The weighting factor, in the first term of Equation 4.2, was calculated for each mode as the square of the error between the measured natural frequency and the natural frequency calculated from the initial model. The weighting function for the second term in Equation 4.2 was set to 0.75. When the particle-swarm-optimization and genetic-algorithm methods were implemented for finite-element-model updating, the results shown in Table 4.2 were obtained.

On implementing the particle-swarm optimization for finite-element-model updating, a population of 50; a value for c_1 of 0.05 and a value for c_2 of 0.01, as well as a value for w of 0.002 were used. The particle-swarm optimization and genetic-algorithm methods were both run for 200 generations. An arithmetic crossover with a probability of 40% and a nonuniform mutation of probability of 0.5% were used.

When a modulus of elasticity of 7×10^{10} N m^{-2} was assumed, the error between the first measured natural frequency and that from the initial finite-element model was 1.9%. When the particle-swarm-optimization method was used for finite-element-model updating, this error was reduced to 0.0% but when using the genetic algorithm approach it was increased to 2.7%.

The error between the second measured natural frequency and that from the initial model was 2.2%. When particle-swarm optimization was used, this error was reduced to 1.8% and when using the genetic algorithm it was reduced to 0.3%.

The error of the third natural frequencies between the measured data and the initial finite-element model was 1.2%. When particle-swarm optimization was used this error was reduced to 0.0% but using the genetic algorithm it was reduced to 0.5%.

The error between the fourth measured natural frequency and that from the initial model was 1.4%. When particle-swarm optimization was used, this error was reduced to 0.2% and when using the genetic algorithm it was reduced to 0.0%.

Overall, the particle-swarm-optimization method gave the best results with an average error of 0.5%, calculated over all the four natural frequencies, while the genetic-algorithm method gave an average error of 0.9%. On average, both methods were improvements when compared to the average error between the initial finite-element model and the measured data calculated over four modes of 1.7%. These error results are also shown in Figure 4.3.

The updated models implemented were also validated on the mode shapes they predicted. To make this assessment possible the MAC was used. The results are shown in Table 4.3. The mean of the diagonal of the MAC vector was used to compare the mode shapes predicted by the updated and initial finite-element models with the measured mode shapes. The average value of 1.0 indicates that the mode shapes are properly correlated.

The average MAC calculated between the mode shapes from an initial finite-element model and the measured mode shapes was 0.9986. When the average MAC was calculated between the measured data and data obtained from the updated finite-element models, it was observed that the particle-swarm optimization and genetic-algorithm-updated finite-element models both gave improved averages for the diagonals of the MAC matrices of 0.9989.

Table 4.2 Results showing measured frequencies of the beam, the initial frequencies and the frequencies obtained when the finite-element model of a beam is updated using the NM and GA

Modes	Measured frequency (Hz)	Initial frequency (Hz)	Frequencies from the PSO updated model (Hz)	Frequencies from the GA updated model (Hz)
1	41.5	42.3	41.5	40.4
2	114.5	117.0	112.4	114.2
3	224.5	227.3	224.4	223.3
4	371.6	376.9	370.9	371.5

Table 4.3 Beam results showing the MAC calculated between measured mode shapes and the initial finite-element model, the NM updated finite-element model and the PSO updated finite-element model

Method	MAC
Initial model	0.9986
PSO	0.9989
GA	0.9989

4.6 Example 2: Unsymmetrical H-shaped Structure

The unsymmetrical H-shaped aluminum structure shown in Chapter 2 was used to validate the presented method further. This structure was also used by Marwala and Heyns (1998) and by Marwala (1997). The structure had three thin cuts of 1 mm that went half-way through the cross-section of the beam. These cuts were introduced to elements 3, 4 and 5. The structure with these cuts was used so that the initial finite-element model gave data that were far from the measured data and that would thereby test the proposed procedure on a difficult finite-element-model updating problem. The structure was suspended using elastic rubber bands.

The structure was excited using an electromagnetic shaker and the response was measured using an accelerometer. The structure was divided into 12 elements. It was excited at the position indicated and acceleration was measured at 15 positions. The tested structure was freely suspended, and a set of 15 frequency-response functions were calculated. A roving accelerometer was used for the testing. The mass of the accelerometer was found to be negligible compared to the mass of the structure.

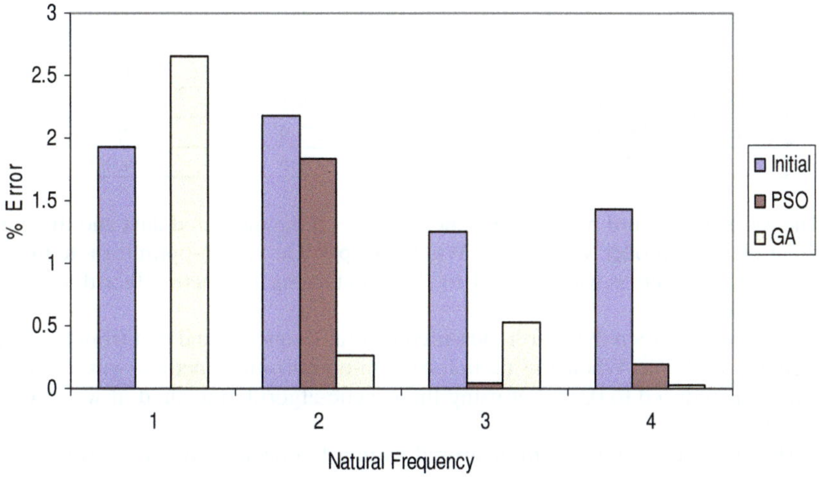

Figure 4.3 Results showing the errors between measurements and the finite-element model

As in the previous example, the finite-element model was constructed using the structural dynamics toolbox (Balmès, 1997) using Euler–Bernoulli beam elements. The finite-element model contained 12 elements. As in the previous example, the moduli of elasticity of these elements used as updating parameters were restricted to fall in the interval 6×10^{10} to 8×10^{10} N m^{-2}.

The particle-swarm-optimization and genetic-algorithm methods were implemented as in the previous example. When the particle-swarm-optimization and genetic-algorithm methods were used for finite-element-model updating, the results shown in Table 4.4 were obtained. Table 4.4 shows the measured natural frequencies, initial natural frequencies and natural frequencies obtained by the particle-swarm optimization and genetic-algorithm-updated finite-element models. When a modulus of elasticity of 7×10^{10} N m^{-2} was assumed, the error between the first measured natural frequency and that from the initial finite-element model was 4.3%. When the particle-swarm-optimization method was used for finite-element-model updating, this error was reduced to 0%. Also, the genetic-algorithm approach reduced this error to 0%.

The error between the second measured natural frequency and that from the initial model was 8.4%. When particle-swarm optimization was used, this error was reduced to 0.4% and the genetic algorithm reduced the error to 2.4%.

Table 4.4 Results from an unsymmetrical H-shaped structure showing measured frequencies, the initial frequencies and the frequencies obtained when the finite-element model was updated using the PSO and GA methods

Modes	Measured frequency (Hz)	Initial frequency (Hz)	Frequencies from the PSO updated model (Hz)	Frequencies from the GA updated model (Hz)
1	53.9	56.2	53.9	53.9
2	117.3	127.1	117.8	120.1
3	208.4	228.4	208.5	211.3
4	254.0	263.4	253.9	253.4
5	445.1	452.4	438.5	438.6

The error of the third natural frequencies between the measured data and the initial finite-element model was 9.6%. When the particle-swarm-optimization method was used, this error was reduced to 0.1% but using a genetic algorithm, it was reduced to 1.4%.

The error between the fourth measured natural frequency and that from the initial model was 3.7%. When the particle-swarm-optimization method was used, this error was reduced to 0.0% but using the genetic-algorithm method, it was reduced to 0.2%.

The error between the fifth measured natural frequency and that from the initial model was 1.6%. When the particle-swarm-optimization method was used, this error was reduced to 1.5% but the genetic-algorithm method increased it to 1.4%.

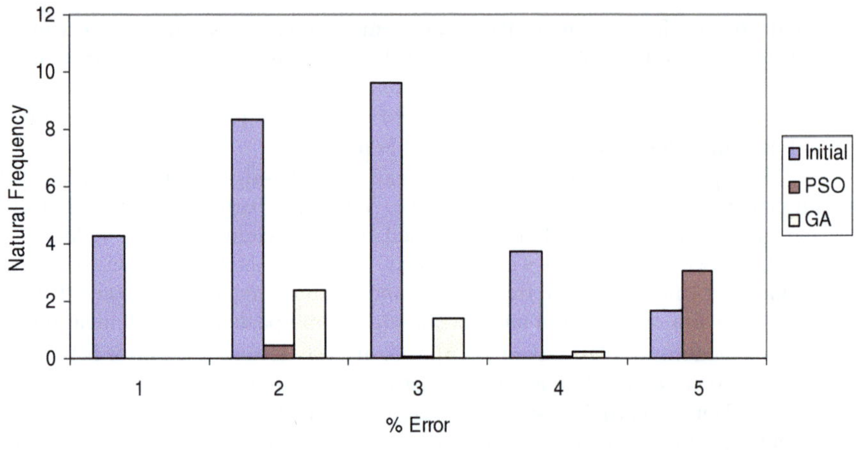

Figure 4.4 Results showing the errors between measurements and finite-element model

Overall, the particle-swarm-optimization method gave the best results with an average error calculated over all the five natural frequencies of 0.4%, while the genetic-algorithm method gave an average error of 1.6%. On average, both methods improved when compared to the average error between the initial finite-element model and the natural frequencies, which was 5.5%. These error results are shown in Figure 4.4.

As in the previous example, the updated models implemented were validated on the mode shapes they predicted using the modal assurance criterion and the results are shown in Table 4.5. Table 4.5 shows that the particle-swarm-optimization and genetic-algorithm-updated finite-element models gave improved averages for the diagonals of the MAC matrices of 0.8434 and 0.8419, respectively.

Table 4.5 Results for the unsymmetrical H-shaped structure, showing measured frequencies, the initial frequencies and the frequencies obtained when the finite-element model is updated using the PSO and GA methods

Method	MAC
Initial model	0.8394
PSO	0.8434
GA	0.8419

4.7 Conclusion

In this chapter, the particle-swarm-optimization method and a genetic-algorithm technique were implemented for finite-element-model updating. When these techniques were tested on finite-element-model updating of a simple beam and an unsymmetrical H-shaped structure, it was observed, on average, that the particle-swarm-optimization method gave more accurate results than a genetic algorithm.

4.8 Future Work

This chapter introduced particle swarm optimization for finite-element-model updating. For further work, methods of designing the entire finite-element-updating–particle-swarm-optimization system for maximum efficiency (accuracy of measured-data estimation) should be identified. Another issue that needs to be addressed is how to determine the theoretical relationship between the number of updating parameters and the accuracy of the finite-element model, given the level of errors in the measured data when particle-swarm optimization is used. There are many variations of particle-swarm-optimization methods. In the future, the relationship between the type of particle-swarm optimization and the effectiveness of the particle-swarm-optimization-based updating process should be identified.

References

Allemang RJ, Brown DL (1982) A Correlation Coefficient for Modal Vector Analysis. In: Proc of the 1st Int Modal Anal Conf:01–18

Arifovic J, Gençay R (2001) Using Genetic Algorithms to Select Architecture of a Feedforward Artificial Neural Network. Physica A: Statistical Mech and its App 289:574–594

Arumugam MS, Rao MVC (2008) Molecular Docking with Multi-objective Particle Swarm Optimization. Appl Soft Comput 8:666–675

Arya LD, Choube SC, Shrivastava M, Kothari DP (2007) The Design of Neuro-fuzzy Networks Using Particle Swarm Optimization and Recursive Singular Value Decomposition. Neurocomputing 71:297–310

Balmès E (1997) Structural Dynamics Toolbox User's Manual Version 2.1. Scientific Software Group, Sèvres, France

Berlinet A, Roland C (2008) A Novel Particle Swarm Optimization Algorithm for Permutation Flow-shop Scheduling to Minimize Makespan. Chaos, Solitons & Fractals 35:851–861

Brits R, Engelbrecht AP, van den Bergh F (2007) Locating Multiple Optima Using Particle Swarm Optimization. Appl Math and Comput 189:1859–1883

Dindar ZA (2004) Artificial Neural Networks Applied to Option Pricing. Unpublished Master Thesis, University of the Witwatersrand, Johannesburg

Dindar ZA, Marwala T (2004) Option Pricing Using a Committee of Neural Networks and Optimized Networks. In: Proc of the 2004 IEEE Int Conf on Syst, Man and Cybern 1:434–438

Engelbrecht AP (2005) Fundamentals of Computational Swarm Intelligence. Wiley

Ewins DJ (1995) Modal Testing: Theory and Practice.Research Studies Press, Letchworth

Friswell MI, Mottershead JE (1995) Finite Element Model Updating in Structural Dynamics. Kluwer Academic Publishers Group, Norwell

Giffler B, Thompson GL (1960) Algorithms for Solving Production Scheduling Problems. Oper Res 8:487–503

Guerra FA, dos S. Coelho L (2008) A Particle Swarm Optimization Approach to Nonlinear Rational Filter Modeling. Expert Syst with Appl 34:1194–1199

Hassan R, Cohani B, de Weck O (2005) A Comparison of Particle Swarm Optimization and Genetic Algorithm. AIAA Paper 2005-1897

Heppner F, Grenander U (1990) A Stochastic Non-linear Model for Coordinated Bird Flocks. In: Krasner S (ed) The Ubiquity of Chaos, 1st edn. Washington DC: AAAS Publications

Hua XG, Ni YQ, Ko JM (2009) Adaptive Regularization Parameter Optimization in Output-error-based Finite Element Model Updating. Mech Syst and Signal Process 23:563–579

Jarboui B, Damak N, Siarry P, Rebai A (2008) The Landscape Adaptive Particle Swarm Optimizer. Appl Soft Comput 8:295–304

Jiang Y, Hu T, Huang C, Wu X (2007) Particle Swarm Optimization Based on Dynamic Niche Technology with Applications to Conceptual Design. Adv in Eng Softw 38:668–676

Janson S, Merkle D, Middendorf M (2007) Strength Design of Composite Beam Using Gradient and Particle Swarm Optimization. Compos Struct 81:471–479

Kathiravan R, Ganguli R (2007) Particle Swarm Optimization for Determining Shortest Distance to Voltage Collapse. Int J of Electr Power & Energy Systems 29:796–802

Kennedy JE, Eberhart RC (1995) Particle Swarm Optimization. In: Proc of the IEEE Int Conf on Neural Netw:942–1948

Kubalík J, Lazanský J (1999) Genetic Algorithms and their Testing. AIP Conf Proc 465:217–229

Levin RI, Lieven NAJ (1998) Dynamic Finite Element Model Updating Using Simulated Annealing and Genetic Algorithms. Mech Syst and Signal Process 12: 91–120

Lian Z, Gu X, Jiao B (2008) Multi-step Ahead Nonlinear Identification of Lorenz's Chaotic System Using Radial Basis Neural Network with Learning by Clustering and Particle Swarm Optimization. Chaos, Solitons & Fractals 35:967–979

Lin Y, Chang W, Hsieh J (2008) Application of Multi-phase Particle Swarm Optimization Technique to Inverse Radiation Problem. J of Quant Spectrosc and Radiat Transf 109:476–493

Liu X, Liu H, Duan H (2007) Time Series Prediction with Recurrent Neural Networks Trained by a Hybrid PSO–EA Algorithm. Neurocomputing 70:2342–2353

Liu L, Liu W, Cartes DA (2008) Particle Swarm Optimization-based Parameter Identification Applied to Permanent Magnet Synchronous Motors. Eng Appl of Artifici Intell 21:1092–1100

Marwala T (1997) A Multiple Criterion Updating Method for Damage Detection on Structures. Masters Thesis, University of Pretoria.

Marwala T (2003) Control of Fermentation Process Using Bayesian Neural Networks and Genetic Algorithm. In: Proc of the African Control Conf: 449–454

Marwala T (2005) Finite Element Model Updating Using Particle Swarm Optimization. Int J of Eng Simul 6:25–30

Marwala T, Heyns PS (1998) A Multiple Criterion Method for Detecting Damage on Structures, AIAA J 195:1494–1501

Oh S, Pedrycz W (2006) Genetic Optimization-driven Multi-layer Hybrid Fuzzy Neural Networks. Simul Model Pract and Theory 14:597–613

Park BJ, Choi HR, Kim HS (2003) A Hybrid Genetic Algorithm for the Job Shop Scheduling Problems. Comput and Ind Eng 45:597–613

Perez RE, Behdinan K (2007) Particle Swarm Approach for Structural Design Optimization. Comput & Struct 85:1579–1588

Qi, H, Ruan LM, Shi M, An W, Tan HP (2008) A Combinatorial Particle Swarm Optimization for Solving Multi-mode Resource-constrained Project Scheduling Problems. Appl Math and Comput 195:299–308

Ransome T (2006) Automatic Minimisation of Patient Setup Errors in Proton Beam Therapy. Unpublished Master Thesis, University of the Witwatersrand, Johannesburg.

Ransome TM, Rubin DM, Marwala T, de Kok EA (2005) Optimising the Verification of Patient Positioning in Proton Beam Therapy. In: Proc of the IEEE 3rd Int Conf on Comput Cybern:279–284

Reynolds CW (1987) Flocks, Herds and Schools: A Distributed Behavioral Model. Computer Graphics 2:25–34

Schlune H, Plos M, Gylltoft K (2009) Improved Bridge Evaluation Through Finite Element Model Updating Using Static and Dynamic Measurements. Eng Struct 31:1477–1485

Sha DY, Hsu C (2006) A Hybrid Particle Swarm Optimization for Job Shop Scheduling Problem. Comput and Ind Eng 51:791–808

Thakker RA, Patil MB, Anil KG (2009) Parameter Extraction for PSP MOSFET Model Using Hierarchical Particle Swarm Optimization. Eng Appl of Artifici Intell 22:317–328

Tu Z, Lu Y (2008) FE Model Updating Using Artificial Boundary Conditions with Genetic Algorithms. Comput & Struct 86:714–727

Weber B, Paultre P, Proulx J (2009) Consistent Regularization of Nonlinear Model Updating for Damage Identification. Mech Syst and Signal Process 23:1965–1985

Yisu J, Knowles J, Hongmei L, Yizeng L, Kell DB (2008) On the Improved Performances of the Particle Swarm Optimization Algorithms with Adaptive Parameters, Cross-over Operators and Root Mean Square (RMS) Variants for Computing Optimal Control of a Class of Hybrid Systems. Appl Soft Comput 8:324–336

Zhang Y, Gallipoli D, Augarde CE (2009) Simulation-based Calibration of Geotechnical Parameters Using Parallel Hybrid Moving Boundary Particle Swarm Optimization. Comput and Geotech 36:604–615

Chapter 5

Finite-element-model Updating Using Simulated Annealing

Abstract. This chapter implements simulated annealing (SA) for updating of a finite-element model using vibration data. This method was tested on a simple beam and an unsymmetrical H-shaped structure and was compared to a method that used the particle-swarm-optimization method (PSO). It was observed that, on average, the particle-swarm-optimization method gives more accurately updated finite elements than the simulated-annealing method. This is mainly due to the simplicity of its implementation.

Keywords: simulated annealing, particle-swarm optimization, cooling schedule, Markov chain Monte Carlo, transition probabilities, Metropolis algorithm

5.1 Introduction

As described in earlier chapters, the aim of finite-element-model updating is to tune the finite-element model to reflect better the measured data taken from the physical structure being modeled (Friswell and Mottershead, 1995).

Chapter 2 presented the Nelder–Mead simplex and the Broyden–Fletcher–Goldfarb–Shanno (BFGS) method for finite-element-model updating. It was observed, on average, that the Nelder–Mead simplex method gave more accurate results than the BFGS.

In Chapter 3, the genetic-algorithm (GA) method was introduced for finite-element model updating. This method was tested with finite element model updating of a simple beam and an unsymmetrical H-shaped structure and compared to a method based on the Nelder–Mead simplex optimization method. It was observed, on average, that the genetic-algorithm method gave a more accurately updated finite-element model than the Nelder–Mead simplex optimization method did.

In Chapter 4, a particle-swarm-optimization (PSO) method was implemented for finite-element-model updating and compared to the genetic-algorithm method.

When tested on a simple beam and an unsymmetrical H-shaped structure it was observed, on average, that the particle-swarm-optimization method gave a more accurately updated finite-element model than the GA method did.

In this chapter, simulated annealing (SA) is introduced for finite-element-model updating and it is compared to the particle-swarm-optimization method, which was identified as being more accurate than genetic algorithms in the previous chapter.

It was Levin and Lieven (1998) who introduced simulated annealing for finite-element-model updating. *Simulated annealing* is an optimization method that is inspired by the natural annealing process where objects such as metals recrystallize when cooled down according to some cooling schedule. It is essentially a Monte Carlo technique that is used to identify an optimal solution. The advantages of simulated annealing for optimization are (van Laarhoven and Aarts, 1997):

- it can handle arbitrary systems and objective functions with any number of design variables;
- statistically, it assures that a global optimum solution is obtained even though this may be after an infinite number of iterations;
- it is simple to code, even for complicated problems, and it is premised on sound statistical grounds; and
- it usually offers a solution that is sufficiently optimal to be of significant use in practice.

Simulated annealing also has the following disadvantages (Salamon *et al.*, 2002):

- If the objective function is expensive to compute recurrently, annealing with a schedule is extremely slow. This is mainly due to the fact that simulated annealing is essentially a Monte Carlo simulation procedure.
- If the objective function is smooth or there are not many local minima, then simulated annealing is not as efficient as other methods such as the conjugate gradient methods (Nocedal and Wright, 2000). Nevertheless, for many problems, the characteristics of the objective function are not known in advance and therefore for a certain class of problems it is not possible to know whether to use the simulated annealing or the conjugate gradient method.
- Simulated annealing is highly dependent on the nature of the problem at hand and it takes advantage of extra information about the system.
- It is difficult to know when the optimal solution has been identified, unlike methods such as the conjugate gradient method, where a second derivative of the objective function can be used to determine whether an optimal solution has been achieved.
- It is often difficult to know which temperature cooling schedule to implement. This concept of a cooling schedule is explained later in this chapter.

5.2 Mathematical Background

In this chapter, the modal properties, *i.e.* natural frequencies and mode shapes are used as a basis for finite-element-model updating using vibration data. In this chapter, it is assumed that the reason why the finite-element model is unable to predict the measured data is that the moduli of elasticity are not correct. Therefore, the aim of the finite-element-model updating is deemed to be a process that is intended to correctly identify the moduli of elasticity that would give the updated finite-element model. To achieve this goal the objective function that was described in Chapters 3 and 4 is minimized. It measures the distance between *measured* natural frequencies and mode shapes and the *calculated* finite-element model natural frequencies and mode shapes:

$$E = \sum_{i=1}^{N} \gamma_i \left(\frac{\omega_i^m - \omega_i^{calc}}{\omega_i^m} \right)^2 + \beta \sum_{i=1}^{N} \left(1 - diag(MAC(\phi_i^{calc}, \phi_i^m))\right) \quad (5.1)$$

Here, m indicates a measured variable, *calc* indicates a calculated variable, ω_i is the ith natural frequency, ϕ_i is the ith mode shape vector, N is the number of modes and γ_i is the weighting factor that measures the relative distance between the initial estimated natural frequency for mode i and the target frequency of the same mode. The parameter β is the weighting function imposed on the mode shapes, MAC is the modal assurance criterion (Allemang and Brown, 1982) and $diag(MAC)_i$ stands for the ith diagonal element of the *MAC* matrix.

The *MAC* is a measure of the correlation between two sets of mode shapes of the same dimensions. In Equation 5.1, the first term has the purpose of ensuring that the natural frequencies estimated by the finite-element model are as close to the measured ones as possible, while the second term ensures that the mode shapes between measurements and the finite-element-model predictions are correlated.

5.3 Simulated Annealing (SA)

Simulated annealing is a Monte Carlo method that is used to identify an optimal solution. It was inspired by the process of annealing where objects, such as metals, recrystallize or liquids freeze. Naderi *et al.* (2009) introduced an improved simulated method for scheduling flow-shops that minimizes the entire completion time and the total tardiness. A technique based on simulated annealing was successfully designed to achieve a trade-off between intensification and diversification mechanisms. A Taguchi scheme was implemented to identify the optimum parameters using the least n number of experiments (*i.e.*, a trial and error process). Examples of successful use of SA follow.

Paya-Zaforteza *et al.* (2009) used simulated annealing for optimizing the design of reinforced-concrete frames. In designing the reinforced-concrete frames, two objective functions: the embedded carbon dioxide emissions and the economic cost were constructed and then optimized. The results obtained from this study indicated that the embedded emissions and the cost were intimately related and that more environmentally friendly solutions than the lowest-cost solution were obtained at a cost increment, which was reasonably suitable in practice.

Kannan and Zacharias (2009) successfully applied simulated annealing to refine and optimize peptide and protein structures. Simulated annealing was found to be considerably more efficient in achieving low-energy structures and structures similar to the experiment when compared to continuous simulations.

Dafflon *et al.* (2009) successfully applied simulated annealing for categorization of heterogeneous aquifers. A novel simulated annealing method was introduced for the integration of high-resolution geophysical and hydrological data. This method was successfully tested on a synthetic data set and then on data collected at the Boise hydro-geophysical research site.

Cretu and Pop (2009) designed acoustic structures using simulated annealing. This work extended the matrix method formalism through using an auxiliary computational method based on simulated annealing to minimize the objective function that was introduced.

Wei-Zhong and Xi-Gang (2009) applied simulated annealing for the optimal synthesis of heat-integrated distillation sequences. An encoding method that used an integer number series was developed to stand for and influence the flow-sheet structure of the system. This method modeled the problem as an implicit, mixed-integer, nonlinear programming problem. Simulated annealing was found to be appropriate for an implicit, mixed-integer, nonlinear programming because when it was implemented it improved the process of solving the problem.

Briant *et al.* (2009) combined the greedy optimization method with multi-criteria simulated annealing for sequencing cars. Their presented method used three criteria in decreasing order of significance and optimized each time based on a criterion without authorizing degradation of the objective from an earlier optimized criterion.

Cosola *et al.* (2008) proposed a universal structure for the identification of hyper-elastic membranes with Moire methods and multi-point simulated annealing. This procedure was used for mechanically characterizing hyper-elastic materials. The characterization procedure was generalized. A multi-level and multi-point simulated annealing that retained the memory of all best records generated in the optimization process was used to discover the unknown material properties.

Lamberti (2008) developed an efficient simulated annealing algorithm for design optimization of truss structures. Results from numerical simulation demonstrated the efficiency and robustness of the proposed method. The method performed well and converged quickly to the optimum solution.

Pedamallu and Ozdamar (2008) studied a hybrid simulated annealing and local search algorithm for constrained optimization. A hybrid simulated annealing method with features that account for both feasibility and optimality characteristics was augmented with a local search method. Numerical experiments demonstrated good results.

Weizhong *et al.* (2008) used a simulated-annealing method for optimal synthesis of distillation. A novel coding technique that used an integer number series was improved to signify and influence the structure of the system. Furthermore, a step-by-step technique was used for column design and cost calculation.

The synthesis problem was formulated using a mixed-integer, nonlinear programming configuration and was successfully solved using an improved simulated-annealing scheme.

Liu *et al.* (2007) successfully used multi-path simulated annealing to solve protein secondary structure elements. A multi-path simulated annealing was used for globally aligning the center and orientation of a particle simultaneously. Sonmez (2007) used simulated annealing for shape optimization of 2-dimensional structures. The results demonstrated that this method showed high reliability even for cases where the whole free boundary was permitted to change.

He and Hwang (2006) successfully implemented damage detection through an adaptive real-parameter hybrid of simulated annealing and genetic algorithm, while Ogura and Sato (2006) implemented an automatic 3-dimensional reconstruction technique using simulated annealing to produce protein projections.

Moita *et al.* (2006) used simulated annealing in the optimal design of vibration control of an adaptive laminated plate, while Chang (2006) applied simulated annealing for demand-side management by optimizing the chiller loading. Other successful applications of simulated annealing include Chang *et al.* (2006) for energy saving, Gomes and Oliveira (2006) for optimal packing problems by combining simulated annealing and linear programming, Bisetty *et al.* (2006) in the study of the pentacyclo-undecane cage amino acid tripeptides as well as McGookin and Murray-Smith (2006) in optimizing submarine maneuvering controllers.

Because of these successful applications of simulated annealing, it is used in this chapter for finite-element-model updating to ensure that the finite-element-model calculated data do predict the measured data better. As indicated before, simulated annealing is inspired by the physical annealing process. In the annealing process, the object, such as a metal, is heated until it is molten and then its temperature is slowly decreased such that the metal, at any given time, is approximately in thermodynamic equilibrium. As the temperature of the object is lowered, the system becomes more ordered and approaches a *frozen* state at $T=0$. If the cooling process is conducted inadequately or the initial temperature of the object is not sufficiently high, the system may become quenched, forming defects or freezing out in metastable states. This indicates that the system is trapped in a local minimum-energy state.

The process that is followed to simulate the annealing process was proposed by Metropolis *et al.* (1953) and it involves choosing an initial state (using the objective function described in Equation 5.1) and temperature, and holding temperature constant, perturbing the initial configuration and computing the error at the new state. If the new error is lower than the old error then the new state is accepted, otherwise if the opposite is the case, then this state is accepted with a low probability. This is essentially a Monte Carlo method.

Simulated annealing replaces a current solution with a "nearby" random solution with a probability that depends on the difference between the corresponding objective function values and the temperature. The temperature decreases throughout the process, so as temperature starts approaching zero, there are fewer random changes in the solution. As is the case in greedy search methods, simulated annealing keeps moving towards the best solution, except that it has the advantage of reversal in fitness. This means it can move to a solution with worse fitness than it has currently achieved, but the advantage of this is that it ensures that the solution is not a local optimum solution, but a global optimum. This is a major advantage that simulated annealing has over other methods but, once again, its drawback is its potentially high computational time. Simulated annealing identifies the global optimum if specified, but it can take an infinite amount of time to achieve this. The probability of accepting the reversal is given by Boltzmann's equation (Bryan *et al.*, 2006):

$$P(\Delta E) = \frac{1}{Z}\exp\left(-\frac{\Delta E}{T}\right) \tag{5.2}$$

Here, ΔE is the difference in error (as indicated in the objective function described in Equation 5.1) between the old and new states. The state indicates the possible updated finite-element models. T is the temperature of the system; Z is a normalization factor that ensures that when the probability function is integrated to infinity it becomes 1.

The rate at which the temperature decreases depends on the cooling schedule chosen. There are many different temperature schedules. Some of the schedules are described in this chapter. On implementing simulated annealing, the first thing to do is to choose its parameters.

5.3.1 Simulated-annealing Parameters

On implementing simulated annealing to the finite-element–model-updating problem, several parameters and choices need to be specified. These are:

- The state space, which is defined in this chapter as a choice of a set of moduli of elasticity that makes a candidate updated finite-element model.
- The objective function described in Equation 5.1. It should be noted that there are many ways in which this objective function could have been constructed, as was demonstrated in Chapter 2.
- The candidate generator mechanism, which is a random-number generator that ensures that a set of moduli elasticity is chosen. In physics, a random walk process is usually preferred.
- The acceptance probability function, which is a process through which a set of elastic moduli may define an updated finite-element model.
- The annealing temperature schedule.

The choice of these parameters has far-reaching consequences on the effectiveness of the simulated annealing method as far as identifying an optimal solution is concerned. However, there is no optimal manner that can be implemented for choosing these parameters that is ideal for all problems and there is no systematic procedure for optimally choosing these parameters for a given problem such as the finite-element-updating problem. Consequently, the choice of these parameters is arbitrary and a method of trial and error is widely used.

5.3.2 Transition Probabilities

When simulated annealing is implemented, a random-walk process is embarked upon for a given temperature. This random-walk process entails moving from one temperature to another. The transition probability is the probability of moving from one state to another. In this chapter, a state represents a given finite-element-updating procedure. This probability is dependent on the present temperature, the order of generating the candidate moves, and the acceptance probability function. In this chapter, a Markov chain Monte Carlo (MCMC) method is used to make a transition from one state to another. The MCMC creates a chain of possible updated finite-element models and accepts or rejects them using the Metropolis algorithm (Meer, 2007).

5.3.3 Monte Carlo Method

The Monte Carlo method is a computational method that uses repeated random sampling to calculate a result (Mathe and Novak, 2007; Akhmatskaya *et al.*, 2009; Ratick and Schwarz, 2009). Monte Carlo methods have been used for simulating physical and mathematical systems. For example, Lai (2009) used the Monte Carlo method for solving matrix and integral problems, while McClarren and Urbatsch (2009) used a modified Monte Carlo method for modeling time-dependent radiative transfer with adaptive material coupling.

Other recent applications of the Monte Carlo method include its use in particle coagulation (Zhao and Zheng, 2009), in diffusion problems (Liu *et al.*, 2009), for the design of radiation detectors (Dunn and Shultis, 2009), for modeling bacterial activities (Oliveira *et al.*, 2009), for vehicle detection (Jia and Zhang, 2009), for modeling the bystander effect (Xia *et al.*, 2009), and for modeling nitrogen absorption (Rahmati and Modarress, 2009).

The Monte Carlo method's reliance on repeated computation and random or pseudo-random numbers necessitates the use of computers. The Monte Carlo method is usually used when it is not possible or not feasible to compute an exact solution using a deterministic algorithm.

5.3.4 Markov Chain Monte Carlo (MCMC)

A Markov chain Monte Carlo (MCMC) method is a process of simulating a chain of states through a random-walk process. It consists of a Markov process and a Monte Carlo simulation (Liesenfeld and Richard, 2008). Jing and Vadakkepat (2009) used a Markov chain Monte Carlo process for tracking of maneuvering objects while Gallagher *et al.* (2009) used the Markov chain Monte Carlo process to identify optimal models, model resolution and model choice for earth-science problems and Curran (2008) applied MCMC in DNA profiling. Other successful applications of the Markov chain Monte Carlo process include its use in environmental modeling (Gauchere *et al.*, 2008), in medical imaging (Jun *et al.*,

2008), in lake-water quality modeling (Malve *et al.*, 2007), in economics (Jacquier *et al.*, 2007) and in statistics (Lombardi, 2007).

Here, a system is now considered whose evolution is described by a stochastic process consisting of random variables $\{x_1, x_2, x_3, \ldots, x_i\}$. A random variable x_i occupies a state x at discrete time i. The list of all possible states that all random variables can possibly occupy is called a state space. If the probability that the system is in state x_{i+1} at time i+1 depends completely on the fact that it was in state x_i at time i, then the random variables $\{x_1, x_2, x_3, \ldots, x_i\}$ form a Markov chain. In the Markov chain Monte Carlo, the transition between states is achieved by adding a random noise (ε) to the current state as follows:

$$x_{i+1} = x_i + \varepsilon \tag{5.3}$$

5.3.5 Acceptance Probability Function: Metropolis Algorithm

When the current state has been achieved, it is either accepted or rejected. In this chapter the acceptance of a state is decided using the Metropolis algorithm (Bedard, 2008; Meyer *et al.*, 2008). This algorithm, which was invented by Metropolis *et al.* (1953), has been used extensively to solve problems of statistical mechanics. Bazavov *et al.* (2009) successfully applied biased Metropolis algorithms for protein simulation. Other applications of the Metropolis algorithms are in nuclear power plants (Sacco *et al.*, 2008), in protein-chain simulation (Tiana *et al.*, 2007), and for the prediction of free Co-Pt nano clusters (Moskovkin and Hou, 2007).

In the Metropolis algorithm, on sampling a stochastic process $\{x_1, x_2, x_3, \ldots, x_i\}$ consisting of random variables, random changes to x are considered and are either accepted or rejected according to the following criterion:

$$\begin{aligned} &if\ E_{new} < E_{old}\ accept\ \ state\ \ (s_{new}) \\ &else \\ &accept\ \ (s_{new})\ with\ \ probabilit\ y \\ &\exp\{-(E_{new} - E_{old})\} \end{aligned} \tag{5.4}$$

5.3.6 Cooling Schedule

A *cooling scheduling* is the process through which the temperature T should be reduced during simulated annealing (De Vicente *et al.*, 2003). Lessons from the physical simulated annealing dictate that the cooling rate should be sufficiently low for the probability distribution of the present state to be close to the thermodynamic equilibrium at all times during the simulation (Das and Chakrabarti, 2005). The time it takes during the simulation for the equilibrium to be restored (which is also called the *relaxation time*) after a change in temperature depends on the shape of the objective function (Equation 5.1), the present temperature and on the candidate generator. The ideal cooling rate is empirically obtained for each problem. The type of simulated annealing that is called *thermodynamic simulated annealing* attempts to sidestep this problem by eliminating the cooling schedule and adjusting the temperature at each step in the simulation, based on the difference in energy between the two states, in accordance to the laws of thermodynamics (Weinberger, 1990). The implementation of simulated annealing is indicated in Figure 5.1.

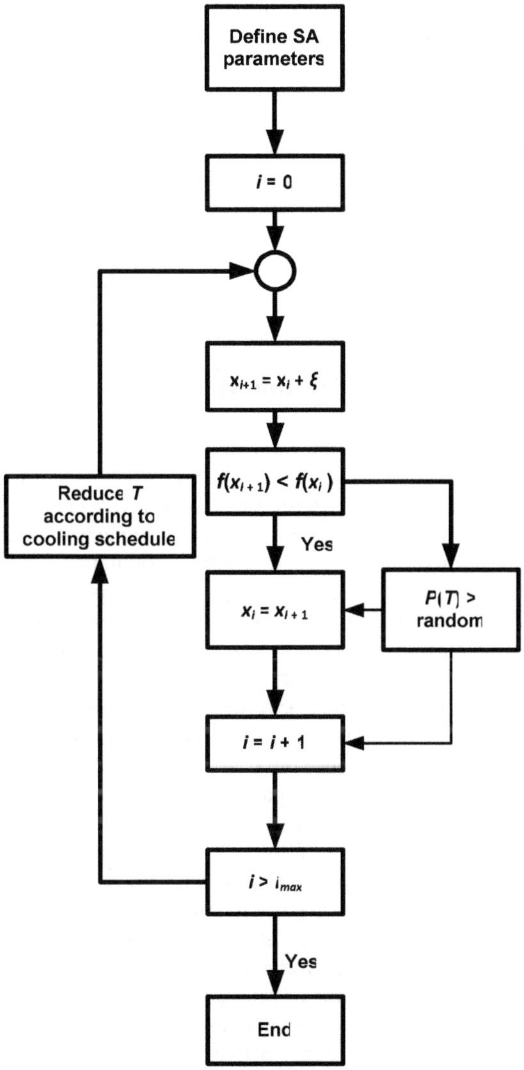

Figure 5.1 The diagram for simulated annealing

The implementation of simulated annealing is indicated in Figure 5.1. The following cooling model is used (Salazar and Toral, 2006):

$$T(i) = \frac{T(i-1)}{1+\sigma} \tag{5.5}$$

Here, $T(i)$ is the current temperature; $T(i-1)$ is the previous temperature and σ is the cooling rate. It must be noted that the precision of the numbers used in the implementation of simulated annealing can have a significant effect on the outcome. A method to improve the computational time of simulated annealing is to implement either very fast simulated reannealing or adaptive simulated annealing (Salazar and Toral, 2006).

This process is repeated such that the sampling statistics for the current temperature is adequate. Then, the temperature is decreased, and the process is repeated until a frozen state is achieved where $T=0$. Simulated annealing was first applied to optimization problems by Kirkpatrick *et al.* (1983). In this chapter, the current state is the current updating solution, the energy equation is the objective function in Equation 5.1, and the ground state is the global optimum solution.

5.4 Particle-swarm-optimization Method

This section compares simulated annealing with particle-swarm optimization (PSO). This is because particle-swarm optimization was found to be more accurate than the genetic-algorithm (GA) method in Chapter 4, which in turn was found to be better than the Nelder–Mead (NM) simplex optimization method in Chapter 3. Particle-swarm optimization is a stochastic, population-based algorithm. As mentioned in Chapter 4, particle-swarm optimization is based on the analogy of flocks of birds, schools of fish, and herds of animals and how they maintain an "evolutionary advantage" by either adapting to their environment, avoiding predators or finding rich sources of food, all by the process of "information sharing" (Marwala, 2005). Particle-swarm optimization initializes by creating a random population of solutions (particles), each solution's fitness is evaluated (according to a fitness function) and the fittest or best solution is noted.

All the solutions in the problem space have their own set of coordinates and these coordinates are associated with the fittest solution found (so far). Another value noted is the best solution found in the neighborhood of the particular solutions or particles. The particle-swarm-optimization method consists of accelerating the velocities of each particle towards the best of what the particles have found and the best of what the swarm has found. The next iteration takes place after all particles have been moved. This behavior mimics the swarming of birds after which the technique was developed.

Table 5.1 shows the difference between simulated annealing and particle-swarm optimization. The main difference between the two is that particle-swarm optimization is an evolutionary, population-based method, while simulated annealing is a statistically based optimization technique inspired by the annealing process. It is clear from Table 5.1 that simulated annealing has relatively fewer parameters to adjust and choose than the particle-swarm-optimization method.

Table 5.1 Operations, types and parameters in the implementation of SA and PSO

Operation	Parameters to select SA	Parameters to select PSO
Representation	Floating point	Floating point
Initialization	Temperature schedule, Initial state	Population size, distribution of the random seed
Operations	Random walk acceptance of samples	Mutation, crossover and reproduction type and probability

5.5 Example 1: Simple Beam

The aluminum beam that was shown in Chapter 2 was used to test the simulated-annealing (SA) method and the particle-swarm-optimization (PSO) method for finite-element-model updating. The beam had the following dimensions: Length = 1.1 m, Width = 29.2 mm, and Thickness = 9.6 mm.

This beam had holes of diameters 5.8 mm located at the centers of elements 2 to 9 and, therefore, was difficult to model. Further details of this beam are reported in Marwala (1997). The beam was tested as being freely suspended using elastic rubber bands. The moduli of elasticity of 11 elements were used as the updating parameters. When the finite-element-model updating was implemented, the moduli of elasticity were restricted to vary from 6×10^{10} to 8×10^{10} N m^{-2}. The weighting factors in the first term of Equation 5.1 were calculated for each mode as the square of the error between the measured natural frequencies and the natural frequencies calculated from the initial model. The weighting function for the second term in Equation 5.1 was set to 0.75. When simulated annealing and particle-swarm optimization were implemented for finite-element-model updating, the results shown in Table 5.2 were obtained.

The particle-swarm optimization was run for 200 generations. On implementing simulated annealing, the scale of the cooling schedule was set to 4 and the number of individual annealing runs was set to 3. When a modulus of elasticity of 7×10^{10} N m^{-2} was assumed, the error between the first measured natural frequency and that from the initial finite-element model was 1.9%. When particle-swarm optimization was used, this error was reduced to 0.0% and using simulated annealing it remained at 1.9%. The error between the second measured natural frequency and that from the initial model was 2.2%. When particle-swarm optimization was used, this error was reduced to 1.8% and using simulated annealing it was reduced to 0.2%.

The error of the third natural frequencies between the measured data and the initial finite-element model was 1.2%. When the particle-swarm-optimization method was used, this error was reduced to 0.0% and using simulated annealing it was reduced to 0.5%. The error between the fourth measured natural frequency and that from the initial model was 1.4%. When the particle-swarm-optimization method was used, this error was reduced to 0.2% and using simulated annealing reduced it to 0.3%. Overall, the particle-swarm-optimization method gave the best results with an average error of 0.5%, calculated over all the four natural frequencies, followed by the simulated annealing with an average error of 0.7%. On average, both methods improved the accuracy of the finite-element model when compared to the average error of 1.7% calculated over four modes between the initial finite-element model and the measured data.

Table 5.2 Results showing measured frequencies of the beam, the initial frequencies and the frequencies obtained when the beam's finite-element model was updated using SA and PSO

Modes	Measured freq (Hz)	Initial freq (Hz)	Frequencies from SA updated model (Hz)	Frequencies from PSO updated model (Hz)
1	41.5	42.3	40.7	41.5
2	114.5	117.0	114.7	112.4
3	224.5	227.3	223.4	224.4
4	371.6	376.9	372.7	370.9

The updated models implemented were also validated on the mode shapes they predicted. To make this assessment possible, the modal assurance criterion (MAC) proposed by Allemang and Brown (1982) was used. The results are shown in Table 5.3.

Table 5.3 Beam results showing the MAC calculated between measured mode shapes and the initial finite-element model, the SA updated finite-element model and the PSO updated finite-element model

Method	MAC
Initial model	0.9986
SA	0.9989
PSO	0.9989

The mean of the diagonal of the modal vector was used to compare the mode shapes predicted by the updated and initial finite-element models with the measured mode shapes. The average value of 1.0 indicates that the mode shapes are properly correlated. The average MAC calculated between the mode shapes from an initial finite-element model and the measured mode shapes was 0.9986. When the average MAC was calculated between the measured data and data obtained from the updated finite-element models, it was observed that particle-swarm-optimization and simulated-annealing-updated finite-element models gave improved averages of 0.9989 for the diagonals of the MAC matrices.

5.6 Example 2: Unsymmetrical H-shaped Structure

The unsymmetrical H-shaped aluminum structure shown in Chapter 2 was also used to validate the proposed method. This structure was previously used by Marwala and Heyns (1998). This structure had three thin cuts of 1 mm that went half-way through the cross-section of the beam. These cuts were introduced to elements 3, 4 and 5. The structure with these cuts was used so that the initial finite-element model gave data that were far from the measured data and, thereby, tested the presented procedure with a difficult finite-element-model updating problem. The structure was suspended using elastic rubber bands. The structure was excited using an electromagnetic shaker and the response was measured using an accelerometer. The structure was divided into 12 elements.

The finite-element model contained 12 elements. As in the previous example, the moduli of elasticity of these elements were used as updating parameters, which were restricted to fall in the interval 6×10^{10} to 8×10^{10} N m^{-2}. The simulated-annealing and particle-swarm-optimization methods were implemented, as was done in the previous example.

When particle-swarm optimization and simulated annealing were used for finite-element-model updating, the results shown in Table 5.4 were obtained. Table 5.4 shows the measured natural frequencies; initial natural frequencies and natural frequencies obtained by the particle-swarm-optimization and simulated-annealing-updated finite-element models.

The error between the first measured natural frequency and that from the initial finite-element model, which was obtained when the modulus of elasticity of 7×10^{10} N m^{-2} was assumed, was 4.3%. When particle-swarm optimization was used, this error was reduced to 0% and simulated annealing reduced this error to 0.2%.

The error between the second measured natural frequency and that from the initial model was 8.4%. When particle-swarm optimization was used, this error was reduced to 0.4% and simulated annealing reduced this error to 1.3%.

The error of the third natural frequencies between the measured data and the initial finite-element model was 9.6%. When particle-swarm optimization was used, this error was reduced to 0.1%, and using the simulated annealing it was reduced to 0.6%.

The error between the fourth measured natural frequency and that from the initial model was 3.7%. When particle-swarm optimization was used, this error was reduced to 0.0% and using simulated annealing reduced it to 0.1%.

The error between the fifth measured natural frequency and that from the initial model was 1.6%. When particle-swarm optimization was used, this error was increased to 1.5% and simulated annealing increased it to 2.1%.

Overall, the particle-swarm-optimization method gave the best results with an average error of 0.4%, calculated over all the five natural frequencies, and then the simulated-annealing method gave an average error of 0.8%. On average, both methods improved the updated finite-element model when compared to the average error between the initial finite-element model and the natural frequencies, which was 5.5%.

Table 5.4 Results from an unsymmetrical H-shaped structure showing measured frequencies, the initial frequencies and the frequencies obtained when the finite-element model is updated using the SA and PSO

Modes	Measured frequency (Hz)	Initial frequency (Hz)	Frequencies from SA updated model (Hz)	Frequencies from PSO updated model (Hz)
1	53.9	56.2	54.0	53.9
2	117.3	127.1	118.8	117.8
3	208.4	228.4	209.7	208.5
4	254.0	263.4	253.8	253.9
5	445.1	452.4	435.8	438.5

As in the previous example, the updated models implemented were validated on the mode shapes they predicted using the MAC and the results are shown in Table 5.5.

Table 5.5 Results for the unsymmetrical H-shaped structure, showing the measured frequencies, initial frequencies and the frequencies obtained when the finite-element model was updated using SA and PSO

Method	MAC
Initial model	0.8394
SA	0.8426
PSO	0.8434

It can be observed from Table 5.5 that the particle-swarm-optimization and simulated-annealing-updated finite-element models gave improved averages for the diagonals of the MAC matrices of 0.8434 and 0.8426, respectively. Another observation from this table is that particle-swarm optimization gives better results than simulated annealing.

5.7 Conclusion

In this chapter, simulated annealing and particle-swarm optimization were implemented for finite-element-model updating. When these techniques were tested on a simple beam and an unsymmetrical H-shaped structure, it was observed that, on average, the particle-swarm-optimization method gave more accurate results than the simulated-annealing method.

5.8 Future Work

This chapter introduced simulated annealing for finite-element-model updating. In further work, methods for designing the entire finite-element-updating-algorithm system for maximum accuracy of measured data estimation should be identified. For further work, other advanced simulated annealing methods such as quantum simulated annealing should be used.

References

Akhmatskaya E, Bou-Rabee N, Reich S (2009) A Comparison of Generalized Hybrid Monte Carlo Methods With and Without Momentum Flip. J of Comput Phys 228:2256–2265

Allemang RJ, Brown DL (1982) A Correlation Coefficient for Modal Vector Analysis. In: Proc of the 1st Int Modal Anal Conf:01–18

Bazavov A, Berg BA, Zhou H (2009) Application of Biased Metropolis Algorithms: From Protons to Proteins. Math and Comput in Simul doi: 10.1016/j.matcom.2009.05.005

Bedard M (2008) Optimal Acceptance Rates for Metropolis Algorithms: Moving beyond 0.234. Stoch Process and their Appl 118: 2198–2222

Bisetty K, Corcho FJ, Canto J, Kruger HG, Perez JJ (2006) Simulated Annealing Study of the Pentacyclo-undecane Cage Amino Acid Tripeptides of the Type [Ac-X-Y-Z-NHMe]. J of Mol Struct: THEOCHEM 759:145–157

Briant O, Naddef D, Mounie G (2009) Greedy Approach and Multi-criteria Simulated Annealing for the Car Sequencing Problem. Eur J of Oper Res 191:993–1003

Bryan K, Cunningham P, Bolshkova N (2006) Application of Simulated Annealing to the Biclustering of Gene Expression Data. IEEE Trans on Inf Technol in Biomed 10519–525

Chang Y (2006) An Innovative Approach for Demand Side Management - Optimal Chiller Loading by Simulated Annealing. Energy 31:1883–1896

Chang Y, Chen W, Lee C, Huang C (2006) Simulated Annealing Based Optimal Chiller Loading for Saving Energy. Energy Convers and Manag 47:2044–2058

Cosola E, Genovese K, Lamberti L, Pappalettere C (2008) A General Framework for Identification of Hyper-elastic Membranes with Moire Techniques and Multi-point Simulated Annealing. Intl J of Solids and Struct 45:6074–6099

Cretu N, Pop M (2009) Acoustic Behavior Design with Simulated Annealing. Comput Mater Sci 44:1312–1318

Curran JM (2008) A MCMC Method for Resolving Two Person Mixtures. Sci and Justice 48:168–177

Dafflon B, Irving J, Holliger K (2009) Simulated-annealing-based Conditional Simulation for the Local-scale Characterization of Heterogeneous Aquifers. J of Appl Geophys 68:60–70

Das A, Chakrabarti BK (2005) Quantum Annealing and Related Optimization Methods. Lect Notes in Phys 679, Springer, Heidelberg

De Vicente J, Lanchares J, Hermida R (2003) Placement by Thermodynamic Simulated Annealing. Phys Lett A 317:415–423

Dunn WL, Shultis JK (2009) Monte Carlo Methods for Design and Analysis of Radiation Detectors. Radiat Phys and Chem 78:852–858

Friswell MI, Mottershead JE (1995) Finite Element Model Updating in Structural Dynamics. Kluwer Academic Publishers Group, Norwell

Gallagher K, Charvin K, Nielsen S, Sambridge M, Stephenson J (2009) Markov Chain Monte Carlo (MCMC) Sampling Methods to Determine Optimal Models, Model Resolution and Model Choice for Earth Science Problems. Marine and Pet Geol 26:525–535

Gauchere C, Campillo F, Misson L, Guiot J, Boreux JJ (2008) Parameterization of a Process-based Tree-growth Model: Comparison of Optimization. MCMC and Particle Filtering Algorithms. Environ Model and Softw 23:1280–1288

Gomes AM, Oliveira JF (2006) Solving Irregular Strip Packing Problems by Hybridising Simulated Annealing and Linear Programming. Eur J of Oper Res 171:811–829

He R, Hwang S (2006) Damage Detection by an Adaptive Real-parameter Simulated Annealing Genetic Algorithm. Comput and Struct 84:2231–2243

Jacquier E, Johannes M, Polson N (2007) MCMC Maximum Likelihood for Latent State Models. J of Econom 137:615–640

Jia Y, Zhang C (2009) Front-view Vehicle Detection by Markov Chain Monte Carlo Method. Pattern Recognit 42:313–321

Jing L, Vadakkepat P (2009) Interacting MCMC Particle Filter for Tracking Maneuvering Target. Digit Signal Process doi: 10.1016/j.dsp.2009.08.011

Jun SC, George JS, Kim W, Pare-Blagoev J, Plis S, Ranken DM, Schmidt DM (2008) Bayesian Brain Source Imaging Based on Combined MEG/EEG and fMRI Using MCMC. NeuroImage 40:1581–1594

Kannan S, Zacharias M (2009) Simulated Annealing Coupled Replica Exchange Molecular Dynamics–An Efficient Conformational Sampling Method. J of Struct Biology 166:288–294

Kirkpatrick S, Gelatt CD, Vecchi MP (1983) Optimization by Simulated Annealing. Sci, New Ser 220:671–680

Lai Y (2009) Adaptive Monte Carlo Methods for Matrix Equations with Applications. J of Comput and Appl Math 231:705–714

Lamberti L (2008) An Efficient Simulated Annealing Algorithm for Design Optimization of Truss Structures. Comput and Struct 86:1936–1953

Levin RI, Lieven NAJ (1998) Dynamic Finite Element Model Updating Using Simulated Annealing and Genetic Algorithms. Mech Syst and Signal Process 12:91–120

Liesenfeld R, Richard J (2008) Improving MCMC, Using Efficient Importance Sampling. Comput Statistics and Data Anal 53:272–288

Liu X, Jiang W, Jakana J, Chiu W (2007) Averaging Tens to Hundreds of Icosahedral Particle Images to Resolve Protein Secondary Structure Elements Using a Multi-path Simulated Annealing Optimization Algorithm. J of Struct Biology 160:11–27

Liu X, Newsome D, Coppens M (2009) Dynamic Monte Carlo Simulations of Binary Self-diffusion in ZSM-5. Microporous and Mesoporous Mater 125:149–159

Lombardi MJ (2007) Bayesian Inference for [Alpha]-Stable Distributions: A Random Walk MCMC Approach. Comput Statistics and Data Anal 51:2688–2700

Malve O, Laine M, Haario H, Kirkkala T, Sarvala J (2007) Bayesian Modelling of Algal Mass Occurrences - Using Adaptive MCMC Methods with a Lake Water Quality Model. Environ Model and Softw 22:966–977

Marwala T (2005) Finite Element Model Updating Using Particle Swarm Optimization. Int J of Eng Simul 6:25–30

Marwala T (1997) A Multiple Criterion Updating Method for Damage Detection on Structures. Master's Thesis, University of Pretoria

Marwala T, Heyns PS (1998) A Multiple Criterion Method for Detecting Damage on Structures, AIAA J 195:1494–1501

Mathe P, Novak E (2007) Simple Monte Carlo and the Metropolis Algorithm. J of Complex 23:673–696

McClarren RG, Urbatsch TJ (2009) A Modified Implicit Monte Carlo Method for Time-dependent Radiative Transfer with Adaptive Material Coupling. J of Comput Phys 228:5669–5686

McGookin EW, Murray-Smith DJ (2006) Submarine Manoeuvring Controllers' Optimisation Using Simulated Annealing and Genetic Algorithms. Control Eng Pract 14:01–15

Meer K (2007) Simulated Annealing versus Metropolis for a TSP Instance. Inf Process Lett 104:216–219

Metropolis N, Rosenbluth A, Rosenbluth M (1953) A. Teller, and E. Teller, Equation of State Calculations by Fast Computing Machines. The J of Chem Phys 21:1087–1092

Meyer R, Cai B, Perron F (2008) Adaptive Rejection Metropolis Sampling Using Lagrange Interpolation Polynomials of Degree 2. Comput Statistics and Data Anal 52:3408–3423

Moita JMS, Correia VMF, Martins PG, Soares CMM, Soares CAM (2006) Optimal Design in Vibration Control of Adaptive Structures Using a Simulated Annealing Algorithm. Compos Struct 75:79–87

Moskovkin P, Hou M (2007) Metropolis Monte Carlo Predictions of Free Co-Pt Nanoclusters. J of Alloy and Compd 434–435:550–554

Naderi B, Zandieh M, Khaleghi A, Balagh G, Roshanaei V (2009) An Improved Simulated Annealing for Hybrid Flowshops with Sequence-dependent Setup and Transportation Times to Minimize Total Completion Time and Total Tardiness. Expert Syst with Appl 36:9625–9633

Nocedal J, Wright S (2000) Numerical Optimization. Springer, Heidelberg

Ogura T, Sato C (2006) A Fully Automatic 3D Reconstruction Method Using Simulated Annealing Enables Accurate Posterioric Angular Assignment of Protein Projections. J of Struct Biology 156:371–386

Oliveira RG, Schneck E, Quinn BE, Konovalov OV, Brandenburg K, Seydel U, Gill T, Hanna CB, Pink DA, Tanaka M (2009) Physical Mechanisms of Bacterial Survival Revealed by Combined Grazing-incidence X-ray Scattering and Monte Carlo Simulation. Comptes Rendus Chimie 12:209–217

Paya-Zaforteza I, Yepes V, Hospitaler A, Gonzalez-Vidosa F (2009) CO2-Optimization of Reinforced Concrete Frames by Simulated Annealing. Eng Struct 31:1501–1508

Pedamallu CS, Ozdamar L (2008) Investigating a Hybrid Simulated Annealing and Local Search Algorithm for Constrained Optimization. Eur J of Oper Res 185:1230–1245

Rahmati M, Modarress H (2009) Nitrogen Adsorption on Nanoporous Zeolites Studied by Grand Canonical Monte Carlo Simulation. J of Mol Struct: THEOCHEM 901:110–116

Ratick S, Schwarz G (2009) Monte Carlo Simulation. In: Kitchin R, Thrift N (ed) International Encyclopedia of Human Geography, Elsevier, Oxford

Sacco WF, Lapa CMF, Pereira CMNA, Filho HA (2008) A Metropolis Algorithm Applied to a Nuclear Power Plant Auxiliary Feedwater System Surveillance Tests Policy Optimization. Prog in Nucl Energy 50:15–21

Salamon P, Sibani P, Frost R (2002) Facts, Conjectures, and Improvements for Simulated Annealing (SIAM Monographs on Mathematical Modeling and Computation). Society for Industrial and Applied Mathematic Publishers, Philadelphia

Salazar R and Toral R (2006) Simulated Annealing Using Hybrid Monte Carlo. arXiv:cond-mat/9706051

Sonmez FO (2007) Shape Optimization of 2D Structures Using Simulated Annealing. Comput Methods in Appl Mech and Eng 196:3279–3299

Tiana G, Sutto L, Broglia RA (2007) Use of the Metropolis Algorithm to Simulate the Dynamics of Protein Chains. Physica A: Statistical Mech and its Appl 380:241–249

van Laarhoven PJ, Aarts EH (1997) Simulated Annealing: Theory and Applications (Mathematics and Its Applications). Kluwer Academic Publishers, Dordrecht

Weinberger E (1990) Correlated and Uncorrelated Fitness Landscapes and How to Tell the Difference. Biological Cybernet 63:325–336

Weizhong AN, Fengjuan YU, Dong F, Yangdong HU (2008) Simulated Annealing Approach to the Optimal Synthesis of Distillation Column with Intermediate Heat Exchangers. Chin J of Chem Eng 16:30–35

Wei-Zhong A, Xi-Gang Y (2009) A Simulated Annealing-based Approach to the Optimal Synthesis of Heat-integrated Distillation Sequences. Comput and Chem Eng 33:199–212

Xia J, Liu L, Xue J, Wang Y, Wu L (2009) Modeling of Radiation-induced Bystander Effect Using Monte Carlo Methods. Nucl Instrum and Methods in Phys Res Section B: Beam Interact with Mater and Atoms 267:1015–1018

Zhao H, Zheng C (2009) Correcting the Multi-Monte Carlo Method for Particle Coagulation. Powder Technology 193:120–123

Chapter 6

Finite-element-model Updating Using the Response-surface Method

Abstract. This chapter presents the response-surface method for finite-element-model updating. The response-surface method was implemented by approximating the finite-element surface-response equation by a multi-layer perceptron, which is a neural-network technique. The updated parameters of the finite-element model were calculated using a genetic algorithm to optimize the surface-response equation. The presented method is compared to existing methods that use simulated annealing and a genetic algorithm separately, with a full finite-element model for model updating. The presented method was tested on a simple and an unsymmetrical H-shaped structure. It was observed that the presented method gave the updated natural frequencies and mode shapes that were an improvement to the initial finite-element model. In general, the accuracy was of the same order of magnitude as those given by the simulated annealing and a genetic algorithm. Furthermore, it was observed that the response-surface method achieves these results with a computational speed that, on average, was at least twice as fast as a genetic algorithm and a full finite-element model and at least twenty times faster than simulated annealing.

Keywords: multi-layer perceptron, response-surface method, back-propagation method, scaled conjugate gradient, neural network

6.1 Introduction

The aim of this chapter is to introduce the updating of finite-element models using the response-surface-method (RSM). Thus far, the RSM has not been widely used to solve the finite-element-model-updating problem (Marwala, 2004a). This approach to finite-element-model updating will be compared to methods that use simulated annealing, implemented in Chapter 5; particle-swarm optimization, implemented in Chapter 4; and a genetic algorithm, implemented in Chapter 3. Finite-element-model updating methods have been implemented using different types of optimization techniques such as a genetic algorithm and the conjugate gradient method (Friswell and Mottershead, 1995; Marwala, 2002; Marwala and Heyns, 1998; Levin and Lieven, 1998).

In Chapter 5, simulated annealing (SA) and particle-swarm optimization (PSO) were compared via the problem of updating the finite-element model of a simple beam and that of an unsymmetrical H-shaped structure. It was observed, on average, that the particle-swarm-optimization method gave a more accurate updated finite-element model than did simulated annealing. In Chapter 4, the particle-swarm-optimization method was implemented and compared to a genetic algorithm (GA) for the same problem, and it was observed, on average, that the particle-swarm-optimization method gave a more accurately updated finite-element model than did a genetic algorithm. In Chapter 3, a genetic algorithm was implemented and compared to the Nelder–Mead (NM) simplex method. Again, for the same problem it was observed, on average, that the genetic algorithm gave better results than the Nelder–Mead simplex method did.

The response-surface method is an approximate optimization method that looks at various design variables and their responses, with the aim of identifying the combination of design variables that give the best response. In this chapter, the best response is defined as the one that gives the minimum distance between the measured data and the data predicted by the finite-element model. The response-surface method tries to replace the implicit functions of the original design-optimization problem with an approximation model, which is traditionally a polynomial and therefore is less expensive to evaluate.

This makes the response-surface method very useful for finite-element-model updating because optimizing the finite-element model for matching measured data to the finite-element model generated data is a computationally expensive exercise. Furthermore, the calculation of the gradients that are essential when using traditional optimization methods, such as conjugate gradient methods, are computationally expensive and the calculation often encounters numerical problems such as ill-conditioning. The response-surface method tends to be immune to such problems when used for finite-element-model updating. This is largely because the response-surface method solves a crude approximation of the finite-element model rather than the full finite-element model, which is of high-dimensional order. The multi-layer perceptron (MLP) (Bishop, 1996) was used in this chapter to approximate the response equation. The response-surface method is particularly useful for optimizing systems that evolve as a function of time, a situation that is prevalent in model-based fault diagnostics found in the manufacturing sector. To date, the response-surface method has been used extensively to optimize complex models and processes (Edwards and Jutan, 1997; Lee *et al.*, 2002). In summary, the response-surface method was used because of the:

- ease of implementation that includes low computational time;
- suitability of the approach to the manufacturing sector where model-based methods are often used to monitor structures that evolve as a function of time; and
- ability to handle complex optimization problems with computationally expensive objective functions.

In addition, the response-surface method has the following advantages (Box and Wilson, 1951):

- it uses statistical models but even the best statistical model is just an approximation to the real system;
- the models and the parameter values are not known and are subjected to uncertainty; and
- an approximated optimum solution may not be the real optimum because of errors of the estimated model and its estimation of the objective function.

The presented response-surface updating method was tested on a simple beam and an unsymmetrical H-shaped structure.

6.2 Mathematical Background

In this chapter, as in previous chapters, natural frequencies and mode shapes were used for finite-element-model updating. In this chapter, to correctly identify the moduli of elasticity that would give the updated finite-element model, the following objective function, as described in Chapters 3 to 5, to measure the distance between measured natural frequencies and mode shapes, and finite-element model calculated data, was minimized:

$$E = \sum_{i=1}^{N} \gamma_i \left(\frac{\omega_i^m - \omega_i^{calc}}{\omega_i^m} \right)^2 + \beta \sum_{i=1}^{N} \left(1 - diag(MAC(\phi_i^{calc}, \phi_i^m))\right) \quad (6.1)$$

In Equation 6.1 m indicates a measured variable, calc indicates a calculated variable, ω_i is the ith natural frequency, ϕ_i is the ith mode shape vector, N is the number of modes and γ_i is the weighting factor that measures the relative distance between the initial estimated natural frequencies for mode i and the target frequency of the same mode. The parameter β is a weighting function on the mode shapes, MAC is the modal assurance criterion (Allemang and Brown, 1982) and $diag(MAC)_i$ stands for the ith diagonal element of the MAC matrix.

6.3 Response-surface Method (RSM)

The response-surface method is a procedure that operates by generating a response for a given input. In this chapter, the inputs are the parameters to be updated and the response is the error between the measured data and the finite-element-model-generated data. An approximation model of the input parameters and the response, called a response-surface equation, is then constructed. Because of this, the optimization method operates on the surface response. This equation is usually simple and not computationally intensive, as opposed to a full finite-element model. The response-surface method has other advantages, such as the ease of implementation through parallel computation and the ease with which parameter sensitivity can be calculated.

Di Lorenzo *et al.* (2010) developed an optimization strategy for hydro-forming process design by using the response-surface method. They hybridized the steepest-descent technique with a moving least-squares method to achieve the optimal internal pressure curve in the hydro-forming of steel tubes. Kankar *et al.* (2009) successfully applied the response-surface method for fault diagnosis in a rotor-bearing system. The response-surface method was used to estimate the impact of design and operating parameters on the vibration signature of a rotor-bearing system.

Kikuchi and Takayama (2009) used a nonlinear response-surface method to assess the reliability of pharmaceutical products. The response surface method was formulated using multi-variate spline interpolation. To estimate the confidence levels of the optimal formulation, a bootstrap resampling method hybridized with a Kohonen's self-organizing map was used. Zeng *et al.* (2009) used the response-surface method for optimizing the design of roll profiles for cold roll forming. The response-surface method was used to relate the impact of forming an angle, a roll radius on spring-back angle and maximum edge membrane longitudinal strains.

Tarley *et al.* (2009) used the combination of response-surface method and factorial design for chemometric tools design, while Nguyen *et al.* (2009) proposed an adaptive response surface for structural reliability analysis. The structural response was calculated from the finite-element method. Normally, the response surface is constructed from polynomials and the unknown coefficients are approximated from a function numerically defined at fitting points. These fitting points were chosen in a well-thought-out manner to diminish the computational time with no worsening of the accuracy of the polynomial estimation. The presented method successively formed the response surface in an integrative way. The response surface was built by the use of a weighted regression technique to weigh the fitting points accordingly. Wei *et al.* (2009) used the response-surface method for spring-back control of sheet-metal forming. To minimize the objective functions of the spring-back and thickness deformation concurrently, a multi-objective genetic algorithm was implemented to locate the optimal solution.

Sereshti *et al.*, (2009) applied the response-surface method for optimizing a dispersive liquid–liquid micro-extraction of water-soluble components, while Rauf *et al.* (2008) applied the response-surface method for photolytic decolorization and data optimization. Tang *et al.* 2008 used the response-surface method for tool shape optimization. To formulate the surface-response function they used a neural network to overcome the limitation of a quadratic polynomial model in solving nonlinear problems. Hu *et al.* (2008) used an adaptive response surface based on an intelligent sampling method for optimizing of sheet-metal-forming processes. This method was successfully tested on the Rosenbrock function and then applied to metal forming. The results show that this procedure gave good models for extremely nonlinear problems with many parameters.

Cheng *et al.* (2008) proposed a neural-network-based response-surface method for structural reliability analysis. A neural network was implemented to build a response-surface method together with the uniform design method for predicting the failure probability of structures. The technique entailed choosing training data for constructing a neural-network model by using the uniform design method. Hou *et al.* (2007) used the response-surface method, the Taguchi method and genetic algorithm for optimizing parameters in nanoparticle wet milling. The response-surface method was used to identify the relationship between the input parameters and output responses, and the genetic algorithm was used to identify the optimal parameters for a nanoparticle milling process.

Tir and Moulai-Mostefa (2008) used the response-surface method for optimizing oil removal from oily wastewater, while Choorit *et al.* (2008) used the response-surface method for the determination of demineralization efficiency in fermented shrimp shells. Other successful applications of the response-surface method include applications in:

- cold wedge rolling (Lee *et al.*, 2008);
- structural reliability analysis (Gavin and Yau, 2008);
- predicting cyclic freeze–thaw damage in concrete structures (Cho, 2007);
- acoustic analysis of damping structures (Liang *et al.*, 2007);
- reliability analysis of allowable pressure on shallow foundation (Babu and Srivastava, 2007);
- vibro-acoustic analysis and optimization of damping structures (Li and Liang, 2007); and
- technologic parameter optimization of a gas–quenching process (Huiping *et al.*, 2007).

The proposed response-surface method for finite-element model updating consists of these essential components:

- the response-surface approximation equation that identifies the relationship between the updating parameters and the natural frequencies and mode shapes; and
- the optimization procedure that is used to identify the updating parameters that will give the measured natural frequencies and mode shapes.

There are many techniques that have been used for response-surface approximation, such as;

- polynomial approximation (Sacks *et al.*, 1989); and
- neural networks (Varaajan *et al.*, 2000).

In this chapter, a type of neural network called the multi-layer perceptron (MLP) was used as a response-surface approximation equation (Bishop, 1996). Further understanding of the different approaches to response-surface approximation may be found in the literature (Giunta and Watson, 1998; Koch *et al.*, 1999; Jin *et al.*, 2000; Lin *et al.*, 2000; Simpson *et al.*, 2001; Wang, 2003).

In this chapter, a multi-layer perceptron was used because it has been successfully used to explain complicated regression problems. The details of the MLP are described in the next section. The second component of the response-surface method is the optimization of the response surface. There are many types of optimization methods that can be used to optimize the response-surface equation and these include the gradient-based methods (Fletcher, 1987) and evolutionary computation methods (Holland, 1975).

In Chapter 3, a genetic algorithm was used for optimization, while in Chapter 4 the particle-swarm-optimization method was used, and in Chapter 5 simulated annealing was used. In this chapter, the response-surface method was used because of its historically good performance. Gradient-based methods such as the conjugate gradient methods have a shortcoming in identifying local optimum solutions, while evolutionary computing methods such as the genetic algorithm and the particle-swarm-optimization method can better identify a global optimum solution. The manner in which the response-surface method was implemented is shown in Figure 6.1.

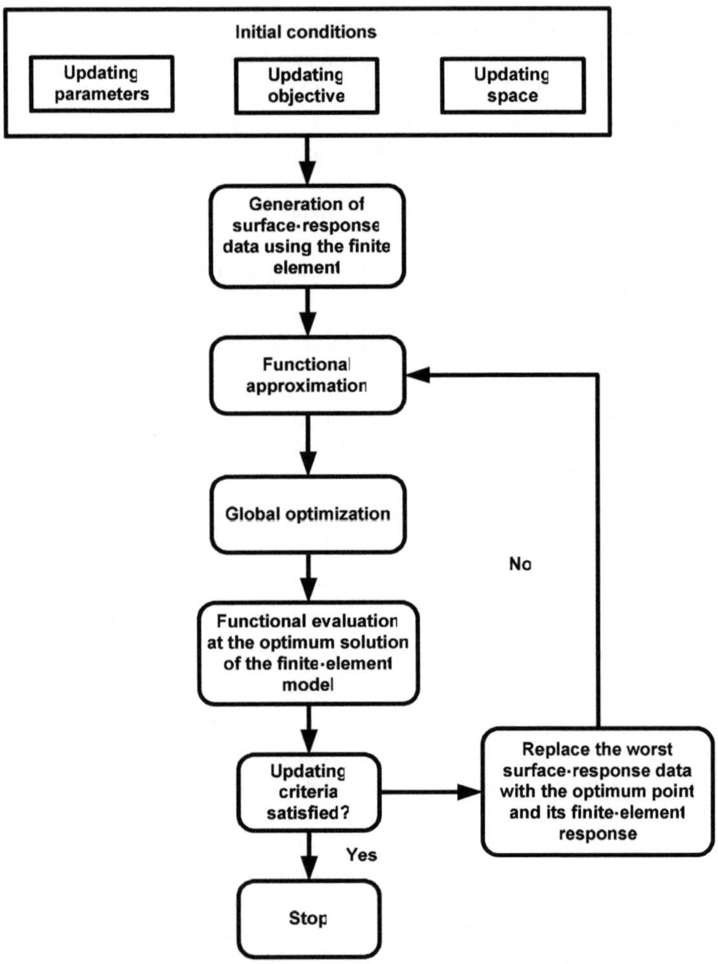

Figure 6.1 Flowchart of the response-surface method

This figure shows that the response-surface method is implemented by following these steps:

- setting the initial conditions, which are updating parameters, updating the objective function, which is represented by Equation 6.1; and
- choosing the updating space such as variable bounds.

The finite-element model is then used to generate sample response-surface data. The MLP is used to approximate the response-surface approximation equation from the data generated in Step 2. Particle-swarm optimization is used to find a global optimum solution. The new optimum solution is used to evaluate the response from the full finite-element model. If the optimum solution does not satisfy the objective, then the new optimum and the corresponding finite-element-model calculated response replaces the candidate with the worst response in the data set generated in Step 2 and then Steps 3 to 5 are repeated. If the objective is satisfied, then the procedure is stopped and the optimum solution becomes the ultimate solution. Step 6 ensures that the simulation always operates in the region of the optimum solution. The next section describes an MLP, which is used for functional approximation.

6.4 Neural Networks

A neural network is an information-processing paradigm that is inspired by the way biological nervous systems, like the human brain, process information. It is a computer-based machine, designed to model the way the brain performs a particular task (Haykin, 1999). It is an exceptionally powerful instrument that has found successful application in the following areas of study:

- mechanical engineering (Marwala and Hunt, 1999; Vilakazi and Marwala, 2006);
- civil engineering (Marwala, 2000; Marwala, 2004b);
- aerospace engineering (Marwala, 2001; Marwala, 2003);
- biomedical engineering (Mohamed *et al.*, 2006; Marwala, 2007);
- finance (Patel and Marwala, 2006); and
- political science (Marwala and Lagazio, 2004).

In this chapter, neural networks are viewed as being generalized regression models that can model the data, which can either be linear or nonlinear. A neural network consists of four main parts (Haykin, 1999). These are:

- the processing units u_j, where each unit has a certain activation level $a_j(t)$ at any point in time;
- weighted interconnections between various processing units. These inter connections determine how the activation of one unit leads to the input for another unit;
- an activation rule, which acts on the set of input signals at a unit to produce a new output signal; and

- the learning rule that specifies how to adjust the weights for a given input or output pair (Haykin, 1999).

Due to their ability to gain meaning from complicated data, neural networks are employed to extract patterns and detect trends that are too complex to be noticed by many other computer techniques (Hassoun, 1995). A trained neural network can be considered as an expert in the category of information it has been given to analyze (Yoon and Peterson, 1990; Lunga and Marwala, 2006). This expert can then be used to provide predictions, given new situations. Because of their ability to adapt to nonlinear data, neural networks have been used to model various nonlinear applications (Haykin, 1999; Hassoun, 1995; Leke *et al.*, 2007).

The configuration of neural processing units and their inter-connections can have a profound impact on the processing capabilities of neural networks (Haykin, 1999). Consequently, many different connections define the flow of data between the input, hidden and output layers. The next section gives details on the architecture of the multi-layer perceptron neural network that was employed in this chapter.

6.4.1 Multi-layer Perceptron (MLP)

A multi-layer perceptron can be defined as a feedforward neural-network model that approximates the relationship between a set of input data and a set of appropriate output data. In this chapter the updating parameters and the objective function are as defined in Equation 6.1.

Its foundation is the standard linear perceptron. It makes use of three or more layers of neurons (nodes) with nonlinear activation functions and is more powerful than the perceptron method. This is because it can distinguish data that are not linearly separable, or are separable by a hyperplane. The multi-layer perceptron has been used to model many complex systems in areas such as mechanical engineering (Marwala, 2001).

The multi-layer perceptron neural network consists of multiple layers of computational units, usually interconnected in a feedforward way (Haykin, 1999; Hassoun, 1995). Each neuron in one layer is directly connected to the neurons of the subsequent layer. A, fully connected, two-layered multi-layer perceptron architecture was used in this chapter. A NETLAB® toolbox (Nabney, 2001) that runs in MATLAB® was used to implement the MLP neural network. Multi-layer perceptron architecture with the two layers shown in Figure 6.2, was used because of the universal approximation theorem, which states that a two-layered architecture is adequate for a multi-layer perceptron and, therefore, it can approximate data of arbitrary complexity (Nabney, 2001; Haykin, 1999).

The network can be mathematically described as follows (Bishop, 1996):

$$y_k = f_{outer}\left(\sum_{j=1}^{M} w_{kj}^{(2)} f_{inner}\left(\sum_{i=1}^{d} w_{ji}^{(1)} x_i + w_{j0}^{(1)}\right) + w_{k0}^{(2)}\right) \tag{6.2}$$

In Equation 6.2 $w_{ji}^{(1)}$ and $w_{ji}^{(2)}$ indicate weights in the first and second layers, respectively, going from input i to hidden unit j; M is the number of hidden units; d is the number of output units; $w_{j0}^{(1)}$ and $w_{k0}^{(2)}$ are the free parameters that indicate the biases for the hidden unit j and the output unit k; $f_{\text{outer}}(\bullet)$ is the logistic activation function; and f_{inner} is the hyperbolic tangent activation function.

The choice of activation allows modeling linear and nonlinear data of any order. These free parameters can be viewed as a mechanism that makes the model actually understand the data. The logistic function is defined as follows (Bishop, 1996):

$$f_{outer}(\nu) = \frac{1}{1+e^{-\nu}} \tag{6.3}$$

The logistic activation function maps the interval (-∞, ∞) onto a (0, 1) interval and can be approximated by a linear function, provided that the magnitude of ν is small. The hyperbolic tangent function is:

$$f_{inner}(\nu) = \tanh(\nu) \tag{6.4}$$

6.4.2 Training the Multi-layer Perceptron

The process of training the neural network consists of identifying the weights in Equation 6.2, given the data. This training process results in a model that embodies the rules and interrelationships that govern the data. As described before, an objective function that represents some distance between the model prediction and the observed target data with the free parameters as unknown variables must be chosen for optimization. Minimizing this cost function, therefore, identifies the free parameters known as weights, given the training data, as depicted in Equation 6.2. An objective function is a mathematical representation of the overall objective of the problem. In this section, the main objective of neural-network training is to identify a set of neural-network weights that map the variables of the finite-element model that are deemed uncertain onto the natural frequencies and mode shapes. If the training data set $D = \{x_k, y_k\}_{k=1}^{N}$ is used and assuming that the targets y are sampled independently given the inputs x_k and the weight parameters, w_{kj}, the objective function, E, may therefore be written as follows, using the sum-of-squares of errors objective function (Bishop, 1996), as defined by Equation 6.5.

In Equation 6.5 N is the index for the training example, k is the index for the output units, $\{x\}$ is the input vector and $\{w\}$ is the weight vector.

$$
\begin{aligned}
E &= \sum_{n=1}^{N}\sum_{k=1}^{K}\{t_{nk} - y_{nk}\}^2 \\
&= \sum_{n=1}^{N}\sum_{k=1}^{K}\{t_{nk} - y_{nk}(\{x\},\{w\})\}^2
\end{aligned}
\tag{6.5}
$$

Before neural-network training is performed, the network architecture needs to be constructed by choosing the number of hidden units, M. If M is too small, the neural network will be insufficiently flexible and will give poor generalization of the data because of high bias. However, if M is too large, the neural network will be too complex and, therefore, be unnecessarily flexible, and will consequently provide poor generalization due to the phenomenon known as over-fitting caused by high variance. The process of choosing an appropriate M is known as model selection and will be discussed in detail in this chapter.

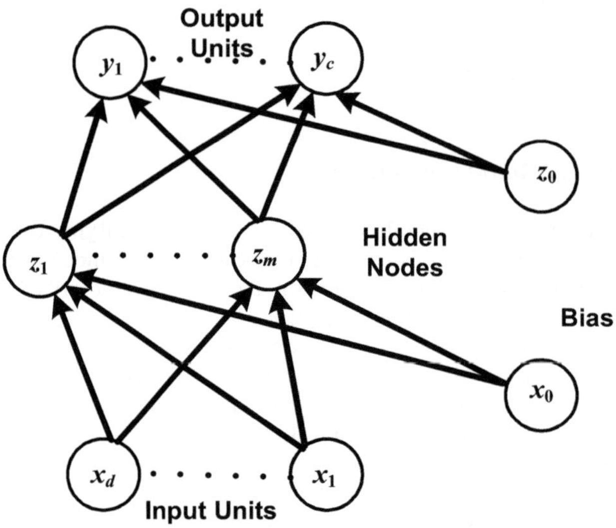

Figure 6.2 A feedforward neural network having two layers of adaptive weights

To train the multi-layer perceptron network using the maximum-likelihood method, as is done in this chapter, a procedure called *back-propagation*, which is the subject of the next section, needs to be implemented. Back-propagation is essentially a method for finding the derivatives of the error, shown in Equation 6.5, with respect to the network weights. This then allows one to implement standard, gradient-based optimization methods to identify the optimal free parameters that can best describe the observed training data.

6.4.3 Back-propagation Method

To identify the network weights, given the training data, an optimization method can be implemented within the context of the maximum-likelihood framework. In general, the weights can be identified using the following iterative method (Werbos, 1974):

$$\{w\}_{i+1} = \{w\}_i - \eta \frac{\partial E}{\partial \{w\}}(\{w\}_i) \tag{6.6}$$

In Equation 6.6 η is the learning rate and {} represents a vector.

The minimization of the objective function, *E,* is achieved by calculating the derivative of the errors in Equations 6.5, with respect to the network weight. The derivative of the error is calculated with respect to the weight that connects the hidden layer to the output layer and may be written using the chain rule as follows (Bishop, 1996):

$$\begin{aligned} \frac{\partial E}{\partial w_{kj}} &= \frac{\partial E}{\partial a_k}\frac{\partial a_k}{\partial w_{kj}} \\ &= \frac{\partial E}{\partial y_k}\frac{\partial y_k}{\partial a_k}\frac{\partial a_k}{\partial w_{kj}} \\ &= \sum_n f'_{outer}(a_k)\frac{\partial E}{\partial y_{nk}} z_j \end{aligned} \tag{6.7}$$

Here, $z_j = f_{inner}(a_j)$ and $a_k = \sum_{j=0}^{M} w_{kj}^{(2)} y_j$.

The derivative of the sum of square cost function in Equation 6.5 may thus be written as follows:

$$\frac{\partial E}{\partial y_{nk}} = t_{nk} - y_{nk} \tag{6.8}$$

Now that it has been determined how to calculate the gradient of the error with respect to the network weights using back-propagation algorithms, Equation 6.6 can then be used to update the network weights using an optimization process until some predefined stopping condition is achieved. If the learning rate in Equation 6.6 is fixed, then this is known as the steepest-descent optimization method (Robbins and Monro, 1951). On the other hand, the steepest-descent method is not computationally efficient and, therefore, an improved method needs to be found. In this chapter the scaled conjugate gradient method is implemented (Møller, 1993), which is the subject of the next section.

6.4.4 Scaled-conjugate-gradient Method

The mechanism in which the free parameters (network weights) are deduced from the data is by the use of some nonlinear optimization method (Mordecai, 2003). In this chapter the scaled conjugate gradient method is used. Before the scaled conjugate gradient method can be described, it is vital to understand how it works. As indicated before, the weight vector that gives the minimum error is achieved by taking successive steps through the weight space as shown in Equation 6.6, until some stopping criterion is attained. Different algorithms choose this learning rate differently. In this section, the gradient-descent method will be discussed, followed by how it is extended to the conjugate gradient method (Hestenes and Stiefel, 1952). For the gradient-descent method, the step size is defined as $-\eta\partial E/\partial w$, where the parameter η is the learning rate and the gradient of the error is calculated using the back-propagation technique, which was described in the previous section.

If the learning rate is sufficiently small, the error value decreases at each successive step until a minimum value for the error between the model prediction and the training target data is obtained. The disadvantage with this approach is that it is computationally expensive when compared to other techniques. For the conjugate gradient method, the quadratic function of error is minimized, at each step over a progressively expanding linear vector space that includes the global minimum of the error (Luenberger, 1984; Fletcher, 1987; Bertsekas, 1995).

For the conjugate gradient procedure, the following steps are followed (Haykin, 1999):

1. Choose the initial weight vector $\{w\}_0$.
2. Calculate the gradient vector $\partial E/\partial\{w\}(\{w\}_0)$.
3. At each step n, use the line search to find $\eta(n)$ that minimizes $E(\eta)$ representing the cost function expressed in terms of η for fixed values of w and $-\partial E/\partial\{w\}(\{w_n\})$.
4. Check that the Euclidean norm of the vector $-\partial E/\partial\{w\}(\{w_n\})$ is sufficiently less than that of $-\partial E/\partial\{w\}(\{w_0\})$.
5. Update the weight vector using Equation 6.6.
6. For w_{n+1} compute the updated gradient $-\partial E/\partial\{w\}(\{w_{n+1}\})$.
7. Use the Polak–Ribiére method to calculate:
8. $\beta(n+1)=\dfrac{\nabla E(\{w\}_{n+1})^T(\nabla E(\{w\}_{n+1})-\nabla E(\{w\}_n))}{\nabla E(\{w\}_n)^T\nabla E(\{w\}_n)}$
9. Update the direction vector
10. $\dfrac{\partial E}{\partial\{w\}}(\{w\}_{n+2})=\dfrac{\partial E}{\partial\{w\}}(\{w\}_{n+1})-\beta(n+1)\dfrac{\partial E}{\partial\{w\}}(\{w\}_n)$.
11. Set $n=n+1$ and go back to Step 3.
12. Stop when the following condition is satisfied:

13. $\frac{\partial E}{\partial\{w\}}(\{w\}_{n+2})=\varepsilon\frac{\partial E}{\partial\{w\}}(\{w\}_{n+1})$, where ε is a small number.

The scaled conjugate gradient method differs from the conjugate gradient method in that it does not involve the line search, described in Step 3 in the previous section. The step size (see Step 3) is calculated directly by using the following formula (Møller, 1993):

$$\eta(n)=2\left(\eta(n)-P^{T}H(n)P+\eta(n)\|P\|^{2}/\|P\|\right)^{2} \tag{6.10}$$

In Equation 6.10 H is the Hessian of the gradient and P is given by:

$$P=\left(\frac{\partial E(n)}{\partial\{w\}}(n)\right) \tag{6.11}$$

The scaled conjugate gradient method was used because it has been found to solve the optimization problems encountered when training a multi-layer perceptron network more computationally efficiently than the gradient-descent and conjugate gradient methods (Bishop, 1996).

6.5 Evolutionary Optimization

In this chapter, the evolutionary programming methods are used for identifying the finite-element-updating parameters. Evolutionary methods for optimization are methods that are inspired by the evolution of natural systems (Sumathi *et al.*, 2008). Evolution is essentially aimed at optimizing some physical system. In this chapter, three evolutionary methods are used:

- the genetic algorithm (Sivanandam and Deepa, 2007);
- particle-swarm optimization (Poli *et al.*, 2008); and
- simulated annealing (Sharma, 2008).

The genetic algorithm was described in Chapter 3 as being inspired by Darwin's theory of natural evolution. It is a simulation of natural evolution where the law of the survival of the fittest is applied to a population of individuals. This natural optimization method was used in this chapter to optimize either the response-surface approximation equation or the error between the finite-element model and the measured data. As in Chapter 3, a genetic algorithm was implemented by generating a population and creating a new population by performing the following procedures:

- crossover;
- mutation; and
- reproduction.

The details of these procedures can be found in Holland (1975) and Goldberg (1989). The *crossover operator* mixes genetic information in the population by cutting pairs of chromosomes at random points along their length and exchanging over the cut sections. This has a potential for joining successful operators together. *Mutation* is a process that introduces new information to a population. *Reproduction* takes successful chromosomes and reproduces them in accordance with their fitness functions. By implementing reproduction, the least-fit members of the population are gradually driven out of the population.

The particle-swarm-optimization method, described in Chapter 4, is an evolutionary programming method that was developed by Kennedy and Eberhart (1995). This procedure was inspired by algorithms that model the "flocking behavior" seen in birds. Researchers in artificial life developed simulations of bird flocking. Heppner and Grenander (1990) conducted simulations where birds were attracted to a roosting area. During the simulations, the birds would start by randomly flying about the space with no particular purpose and instinctively form flocks until one of the birds flew over the roosting area. If the bird's "wish to roost" was greater than the "wish to stay in the flock" the bird pulls away from the flock and lands on the roost. In a simulation, each bird attempts to either remain at the center of other birds or not to run into any of them.

Because birds use simple rules to set their direction and velocity, if a bird pulls away from the flock and lands on the roost, the other birds would also move towards the roost. As more birds notice the roost, they would land there, and this would attract more birds towards the roost, until the entire flock has landed at the roost.

In the context of optimization, the process in which birds find a roost is analogous to finding an optimization process. The means by which a bird that has found the roost causes other birds close by to find the roost also is called *social intelligence*. The particle-swarm-optimization framework effectively states that individuals learn from the successes of their neighbors. So by imitating their behavior it is implemented by finding a balance between searching for a good solution and exploiting someone else's success. If the search for a solution is too limited, the simulation will converge to the first solution encountered, which may be a local optimum position. If the successes of others are not exploited then the simulation will never converge.

Simulated annealing is a Monte Carlo method that was used to investigate the equations of state and frozen states of an n degree-of-freedom system (Metropolis *et al.*, 1953; Kirkpatrick *et al.*, 1983). Simulated annealing was inspired by the process of physical annealing where objects, such as metals, recrystallize or liquids freeze. In the physical annealing process, the object is heated until it is molten, then it is slowly cooled down such that the metal, at any given time, is approximately in thermodynamic equilibrium. As the temperature of the object is lowered, the system becomes more ordered and approaches a *frozen* state. If the cooling process is conducted insufficiently or the initial temperature of the object is not sufficiently high, the system may become quenched, forming defects or freezing out in metastable states. This indicates that the system is trapped in a local minimum energy state that, within the context of optimization, is the local optimum point.

6.6 Example 1: Simple Beam

An aluminum beam shown in Chapter 2 was used to test the response surface method (RSM) for finite-element-model updating. This method is compared to simulated annealing, particle-swarm optimization and a genetic algorithm. The details on this beam can be found in Chapter 2. The moduli of elasticity of 11 elements of this simple beam were used as updating parameters. When the finite-element-model updating was implemented, the moduli of elasticity were restricted to vary from 6×10^{10} to 8×10^{10} N m^{-2}. The weighting factors, in the first term in Equation 6.1, were calculated for each mode as the square of the error between the measured natural frequency and the natural frequency calculated from the initial model. The weighting function for the second term of Equation 6.1 was set to 0.75.

The genetic algorithm was run for a population of 50 and over 200 generations. There are many selection methods and these include roulette-wheel selection (Mohamed *et al.*, 2008), which was used in this chapter. An arithmetic crossover probability of 40% and a nonuniform mutation probability of 0.5% were implemented. On implementing the presented particle-swarm-optimization method for finite-element-model updating the following parameters were chosen: a population of 50; $c_1 = 0.05$ and $c_2 = 0.01$; as well as $w = 0.002$.

The particle-swarm-optimization method was run over 200 generations. On implementing simulated annealing, the scale of the cooling schedule was set to 4 and the number of individual annealing runs was set to 3. On implementing the presented response-surface method, the finite-element model was run 200 times to generate the data for functional approximation. The multi-layer perceptron neural networks implemented had 11 input variables corresponding to the 11 elements in the finite-element model, 8 hidden units and 1 output unit.

As described before, the multi-layer perceptron had a hyperbolic tangent activation function in the hidden layer and a linear activation function in the output layer. The response-surface method functional approximation via the multi-layer perceptron was evaluated 10 times (10 iterations), each time using the particle-swarm-optimization method to calculate the optimum point and evaluating this optimum point on the finite-element-updating model and then storing the previous optimum point in the data set for the current functional approximation. The scaled conjugate gradient method was used to train the multi-layer perceptron, primarily because of its computational efficiency. The initial functional approximation was obtained by training the multi-layer perceptron for 150 training cycles and on a subsequent functional approximation, where the data set had the previous optimum solution added to it, used 5 training cycles. On using the response-surface method, the multi-layer perceptron was only initialized once. When the response-surface method, simulated annealing, particle-swarm-optimization method and genetic algorithm were implemented for finite-element-model updating, the results shown in Table 6.1 were obtained.

If a modulus of elasticity of 7×10^{10} N m^{-2} was assumed, then the error between the first measured natural frequency and that from the initial finite-element model was 1.9%. If a modulus of elasticity of 7×10^{10} N m^{-2} was assumed, the error between the first measured natural frequency and that from the initial finite-element model was 1.9%. When the response-surface method was used for finite-element-model updating, this error was reduced to 1.7%, while using the particle-swarm-optimization method reduced the error to 0%. Using simulated annealing, the error remained at 1.9% and using the genetic-algorithm approach it increased to 2.7%.

The error between the second measured natural frequency and that from the initial model was 2.2%. When the response-surface method was used, this error was reduced to 0.7% while on using the particle-swarm-optimization method it was reduced to 1.8%. Simulated annealing reduced the error to 0.2%, and using the genetic algorithm it was reduced to 0.3%.

The error of the third natural frequencies between the measured data and the initial finite-element model was 1.2%. When the response-surface method was used, this error was reduced to 1.1%, while using particle-swarm optimization reduced it to 0%. Simulated annealing reduced this error to 0.5% and using the genetic algorithm and a full finite-element model reduced it to 0.5%.

Table 6.1 Results showing measured frequencies of the beam, the initial frequencies and the frequencies obtained when the finite-element model of a beam is updated using the RSM, GA, SA and PSO

Modes	1	2	3	4
Measured frequency	41.5	114.5	224.5	371.6
Initial frequency (Hz)	42.3	117.0	227.3	376.9
Frequencies from GA updated model (Hz)	40.4	114.2	223.3	371.5
Frequencies from SA updated model (Hz)	40.7	114.7	223.4	372.7
Frequencies from PSO updated model (Hz)	41.5	112.4	224.4	370.9
Frequencies from RSM updated model (Hz)	40.8	113.7	222.1	370.2

The error between the fourth measured natural frequency and that from the initial model was 1.4%. When the response-surface method was used for finite-element-model updating, this error was reduced to 0.4%, while using particle-swarm optimization reduced it to 0.2%. Simulated annealing reduced the error to 0.3% and using the genetic algorithm and a full finite-element model reduced it to 0%.

Overall, the particle-swarm-optimization method gave the best results with an average error, calculated over all the four natural frequencies, of 0.5%, followed by simulated annealing with an average error of 0.7%, followed by the genetic algorithm with an average error of 0.9% and then the response-surface method with an average error of 1.0%. All these methods, on average, showed improvement when compared to the average error between the initial finite-element model and the measured data, which was 1.7%.

The updated models implemented were also validated on the mode shapes they predicted. To make this assessment possible, the modal assurance criterion (MAC) was used (Allemang and Brown, 1982). The mean of the diagonal of the MAC vector was used to compare the mode shapes predicted by the updated and initial finite-element models to the measured mode shapes. The average MAC calculated between the mode shapes from an initial finite-element model and the measured mode shapes was 0.9986. When the average MAC was calculated between the measured data and data obtained from the updated finite-element models, it was observed that the response-surface method, the particle-swarm-optimization method, simulated annealing and genetic-algorithm-updated-finite-element model gave improved averages for the diagonals of the MAC matrix of 0.9988, 0.9989, 0.9989 and 0.9989, respectively. Therefore, the simulated annealing and genetic algorithm gave the best MAC, which was marginally better than the response-surface method. However, these differences in accuracies of the MAC and natural frequencies were not very significant. The results are summarized in Table 6.2.

The computational time taken to run the complete response-surface method was 44 CPU, while the particle-swarm-optimization method and a full finite-element model took 60 CPU to run. It took simulated annealing and a full finite-element model 900 CPU to run and the genetic algorithm and a full finite-element model took 90 CPU to run. The response-surface method was found to be faster than the particle-swarm-optimization method, which was in turn faster than the genetic algorithm, which was in turn much faster than the simulated annealing method.

Table 6.2 Beam results showing the MAC calculated between measured mode shapes and the initial finite-element model, the GA, SA, PSO and RSM updated finite-element model

Method	MAC
Initial model	0.9986
GA	0.9989
SA	0.9989
PSO	0.9989
RSM	0.9988

6.7 Example 2: Unsymmetrical H-shaped Structure

The unsymmetrical, H-shaped, aluminum structure shown in Chapter 2 was also used to validate the method presented here. The details of the structure were outlined in Chapter 2. The finite-element model contained 12 elements. As in the previous example, the moduli of elasticity of these elements were used as updating parameters that were restricted to fall in the interval 6×10^{10} to 8×10^{10} N m^{-2}. The genetic algorithm, simulated annealing, particle-swarm-optimization method and response-surface method were implemented as in the previous example.

The results obtained in implementing these optimization methods for finite-element-model updating are shown in Table 6.3. The table shows the measured natural frequencies, initial natural frequencies and natural frequencies obtained by the genetic algorithm; simulated annealing, particle-swarm-optimization method and response-surface-method updated finite-element models. When a modulus of elasticity of 7×10^{10} N m^{-2} was assumed, the error between the first measured natural frequency and that from the initial finite-element model was 4.3%.

When the response-surface method was used for finite-element-model updating, this error was reduced to 3.5%. The particle-swarm-optimization method reduced this error to 0%, while simulated annealing reduced this error to 0.2% and the genetic-algorithm approach reduced this error to 0%.

The error between the second measured natural frequency and that of the initial model was 8.4%. When the response-surface method was used, this error was reduced to 1.2%, the particle-swarm-optimization method reduced this error to 0.4%, the simulated annealing reduced this error to 1.3%, and the genetic algorithm reduced this error to 2.4%.

The error in the third natural frequencies between the measured data and the initial finite-element model was 9.6%. When the response-surface method was used, this error was reduced to 0.7%. The particle-swarm-optimization method reduced this error to 0.1%, while using the simulated annealing it was reduced to 0.6% and using the genetic algorithm and a full finite-element model it was reduced to 1.4%.

The error between the fourth measured natural frequency and that of the initial model was 3.7%. When the response-surface method was used, this error was reduced to 0.9%. The particle-swarm-optimization method reduced this error to 0%, while using simulated annealing reduced it to 0.1% and using the genetic algorithm and a full finite-element model it was reduced to 0.2%.

Table 6.3 Results from an unsymmetrical, H-shaped structure showing measured frequencies, the initial frequencies and the frequencies obtained when the finite-element model was updated using the RSM, GA, SA and PSO methods

Modes	1	2	3	4	5
Measured frequency	53.9	117.3	208.4	254.0	445.1
Initial frequency (Hz)	56.2	117.0	127.1	228.4	452.4
Frequencies from GA updated model (Hz)	53.9	120.1	211.3	253.4	438.6
Frequencies from SA updated model (Hz)	54.0	118.8	209.7	253.8	435.8
Frequencies from PSO updated model (Hz)	53.9	117.8	208.5	253.9	438.5
Frequencies from RSM updated model (Hz)	52.0	118.7	209.9	251.6	432.5

The error between the fifth measured natural frequency and that of the initial model was 1.6%. When the response-surface method was used, this error was increased to 2.9%, while the particle-swarm-optimization method reduced this error to 1.5%, simulated annealing increased it to 2.1% and the genetic algorithm changed it to 1.4%.

Overall, the particle-swarm-optimization method gave the best results with an average error of 0.4% calculated over all the five natural frequencies. Then, simulated annealing had an average error of 0.8%, the genetic algorithm had an average error of 1.4% and then the response-surface method had an average error of 1.8%. All four methods, on average, were an improvement on the average error between the initial finite-element model and the natural frequencies, which was 5.5%.

As in the previous example, the updated models implemented were validated on the mode shapes they predicted using the MAC. The response-surface method, particle-swarm-optimization method, simulated-annealing and genetic-algorithm updated finite-element models gave the improved averages for the diagonals of the MAC matrices of 0.8413, 0.8434, 0.8426 and 0.8437, respectively. Therefore, the genetic algorithm gave the best MAC followed by the particle-swarm-optimization method, followed by the simulated annealing and then the response-surface method. These results are summarized in Table 6.4.

The computational load for each method was monitored and it was observed, on average, that the response-surface method was the most computationally efficient method followed by the genetic algorithm, then the particle-swarm-optimization method and finally the simulated annealing.

Table 6.4 Results of the unsymmetrical, H-shaped structure showing the MAC calculated between measured mode shapes and the initial finite-element model, for the GA, SA, PSO and RSM updated finite-element methods

Method	MAC
Initial model	0.8394
GA	0.8437
SA	0.8426
PSO	0.8434
RSM	0.8413

6.8 Conclusion

In this study, the response-surface method, particle-swarm optimization method, simulated annealing and genetic algorithm were implemented for finite-element-model updating. When these techniques were tested on a simple beam and an unsymmetrical H-shaped structure, it was observed on average that the particle-swarm-optimization method gave more accurate results, than the simulated-annealing, then the genetic-algorithm and finally the response-surface method. The response-surface method was found to be more computationally efficient than the other three methods.

6.9 Future Work

This chapter introduced the response-surface method for finite-element-model updating. The response-surface method was constructed using a multi-layer perceptron neural network. For further work, other techniques such as the radial basis function and support vector machines can be used to construct the response-surface equations. Furthermore, methods for designing the entire surface-response method based on a finite-element updating system for maximum accuracy of measured data estimation should be identified.

References

Allemang RJ, Brown DL (1982) A Correlation Coefficient for Modal Vector Analysis. In: Proc of the 1st Intl Modal Anal Conf:01–18

Babu GLS, Srivastava A (2007) Reliability Analysis of Allowable Pressure on Shallow Foundation Using Response Surface Method. Comput and Geotech 34:187–194

Bertsekas DP (1995) Non-linear Programming. Athenas Scientific, Belmont

Bishop CM (1996) Neural Networks for Pattern Recognition. Clarendon, Oxford

Box GEP, Wilson KB (1951) On the Experimental Attainment of Optimum Conditions (with discussion). J of the R Statistical Soc Ser B 1:1–45

Cheng J, Li QS, Xiao R (2008) A New Artificial Neural Network-based Response Surface Method for Structural Reliability Analysis. Probab Eng Mech 23:51– 63

Cho T (2007) Prediction of Cyclic Freeze-thaw Damage in Concrete Structures Based on Response Surface Method. Constr and Build Mater 21:2031–2040

Choorit W, Patthanamanee W, Manurakchinakorn S (2008) Use of Response Surface Method for the Determination of Demineralization Efficiency in Fermented Shrimp Shells. Bioresour Technol 99:6168–6173

Di Lorenzo R, Ingarao G, Chinesta F (2010) Integration of Gradient Based and Response Surface Methods to Develop a Cascade Optimisation Strategy for Y-shaped Tube Hydroforming Process Design. Adv in Eng Softw 41:336–348

Edwards IM, Jutan A (1997) Optimization and Control Using Response Surface Methods. Comput and Chem Eng 4:441–453

Fletcher R (1987) Practical Methods of Optimization. John Wiley, New York

Friswell MI, Mottershead JE (1995) Finite Element Model Updating in Structural Dynamics. Kluwer Academic Publishers Group, Norwell

Gavin HP, Yau SC (2008) High-order Limit State Functions in the Response Surface Method for Structural Reliability Analysis. Struct Saf 30:162–179

Giunta AA, Watson LT (1998) A Comparison of Approximation Modeling Techniques: Polynomial Versus Interpolating Models. AIAA-98-4758, Am Inst of Aeronaut and Astronaut, Inc:392–401

Goldberg DE (1989) Genetic Algorithms in Search, Optimization and Machine Learning. Addison-Wesley, Reading

Hassoun MH (1995) Fundamentals of Artificial Neural Networks. MIT Press, Cambridge

Haykin S (1999) Neural Networks. Prentice-Hall, New Jersey

Heppner F, Grenander U (1990) A Stochastic Non-linear Model for Coordinated Bird Flocks. In: Krasner S (ed) The Ubiquity of Chaos, 1st edn. Washington DC: AAAS Publications

Hestenes MR, Stiefel E (1952) Methods of Conjugate Gradients for Solving Linear Systems. J of Res of the Natl Bur of Stand 6:409–436

Holland J (1975) Adaptation in Natural and Artificial Systems. University of Michigan Press, Ann Arbor

Hou T, Su C, Liu W (2007) Parameters Optimization of a Nano-particle Wet Milling Process Using the Taguchi Method, Response Surface Method and Genetic Algorithm. Powder Technol 173:153–162

Hu W, Yao LG, Hua ZZ (2008) Optimization of Sheet Metal Forming Processes by Adaptive Response Surface Based on Intelligent Sampling Method. J of Mater Process Technol 197:77–88

Huiping L, Guoqun Z, Shanting N, Yiguo L (2007) Technologic Parameter Optimization of Gas Quenching Process Using Response Surface Method. Comput Mater Sci 38:561–570

Jin R, Chen W, Simpson T (2000) Comparative Studies of Metamodeling Techniques Under Multiple Modeling Criteria. 8th AIAA/NASA/USAF/ISSMO Symp on Multidiscip Anal and Optim

Kankar PK, Harsha SP, Kumar P, Sharma SC (2009) Fault Diagnosis of a Rotor Bearing System Using Response Surface Method. Eur J of Mech - A/Solids 28:841–857

Kennedy JE, Eberhart RC (1995) Particle Swarm Optimization. In: Proc of the IEEE Intl Conf on Neural Netw:942–1948

Kikuchi S, Takayama K (2009) Reliability Assessment for the Optimal Formulations of Pharmaceutical Products Predicted by a Nonlinear Response Surface Method. Intl J of Pharm 374:5–11

Kirkpatrick S, Gelatt CD, Vecchi MP (1983) Optimization by Simulated Annealing. Sci 220:671–680

Koch PN, Simpson TW, Allen JK, Mistree F (1999) Statistical Approximations for Multidisciplinary Design Optimization: The Problem of Size. J of Aircr 36:275–286

Lee SH, Kim HY, Oh SI (2002) Cylindrical Tube Optimization Using Response Surface Method Based on Stochastic Process. J of Mater Process Technol 20:490–496

Lee HW, Lee GA, Yoon DJ, Choi S, Na KH, Hwang MY (2008) Optimization of Design Parameters Using a Response Surface Method in a Cold Cross-wedge Rolling. J of Mater Process Technol 201:112–117

Leke B, Marwala T, Tettey T (2007) Using Inverse Neural Networks for HIV Adaptive Control. Intl J of Comput Intell Res 3:11–15

Levin RI, Lieven NAJ (1998) Dynamic Finite Element Model Updating Using Simulated Annealing and Genetic Algorithms. Mech Syst and Signal Process 12: 91–120

Li Z, Liang X (2007) Vibro-acoustic Analysis and Optimization of Damping Structure with Response Surface Method. Mater and Des 28:199–2007

Liang X, Lin Z, Zhu P (2007) Acoustic Analysis of Damping Structure with Response Surface Method. Appl Acoust 68:1036–1053

Lin Y, Krishnapur K, Allen JK, Mistree F (2000) Robust Concept Exploration in Engineering Design: Metamodeling Techniques and Goal Formulations. In: Proc of the 2000 ASME Des Eng Tech Conf, DETC2000/DAC-14283

Luenberger DG (1984) Linear and Non-linear Programming. Addison-Wesley, Reading

Lunga D, Marwala T (2006) On-line Forecasting of Stock Market Movement Direction Using the Improved Incremental Algorithm. Lect Notes in Comput Sci 4234:440–449

Marwala T (2000) On Damage Identification Using a Committee of Neural Networks. J of Eng Mech 126:43–50

Marwala T (2001) On Fault Identification Using Pseudo-modal Energies and Modal Properties. Am Inst of Aeronaut and Astronaut J 39:1608–1617

Marwala T (2002) Finite Element Updating Using Wavelet Data and Genetic Algorithm. AIAA J of Aircr 39:709 711

Marwala T (2003) Fault Classification Using Pseudo Modal Energies and Neural Networks. Am Inst of Aeronaut and Astronaut J 41:82–89

Marwala T (2004a) Finite Element Model Updating Using Response Surface Method. 45th AIAA/ASME/ASCE/AHS/ASC Structures, Structural Dyn & Mater Conf: Paper AIAA-2004-2005

Marwala T (2004b) Fault Classification Using Pseudo Modal Energies and Probabilistic Neural Networks. J of Eng Mech 130:1346–1355

Marwala T (2007) Bayesian Training of Neural Network Using Genetic Programming. Pattern Recognit Lett 28:1452–1458

Marwala T, Heyns PS (1998) A Multiple Criterion Method for Detecting Damage on Structures, AIAA J 195:1494–1501

Marwala T, Hunt HEM (1999) Fault Identification Using Finite Element Models and Neural Networks. Mech Syst and Signal Process 13:475–490

Marwala T, Lagazio M (2004) Modelling and Controlling Interstate Conflict. In: Proc of the IEEE Intl Joint Conf on Neural Netw:1233–1238

Metropolis N, Rosenbluth AW, Rosenbluth MN, Teller AH, Teller E (1953) Equations of State Calculations by Fast Computing Machines. J of Chem Phys 21:1087–1092

Mohamed AK, Nelwamondo FV and Marwala T (2008) Estimation of Missing Data: Neural Networks, Principal Component Analysis and Genetic Algorithms. In: Proc of the 12th World Multi-Conf on Syst, Cybern and Inform:36–41

Mohamed N, Rubin D and Marwala T (2006) Detection of Epileptiform Activity in Human EEG Signals Using Bayesian Neural Networks. Neural Inf Process – Lett and Rev 10:1–10

Mordecai A (2003) Non-linear Programming: Analysis and Methods. Dover Publishing, New York

Møller AF (1993) A Scaled Conjugate Gradient Algorithm for Fast Supervised Learning. Neural Networks 6:525–533

Nabney IT (2001) Netlab: Algorithms for Pattern Recognition. Springer-Verlag, Heidelberg

Nguyen XS, Sellier A, Duprat F, Pons G (2009) Adaptive Response Surface Method Based on a Double Weighted Regression Technique. Probab Eng Mech 24:135–143

Patel P, Marwala T (2006) Neural Networks, Fuzzy Inference Systems and Adaptive-neuro Fuzzy Inference Systems for Financial Decision Making. Lect Notes in Comput Sci 4234:430–439

Poli R, Kennedy J, Blackwell T, Freitas A (2008) Particle Swarms: The Second Decade. Hindawi Publishing

Rauf MA, Marzouki N, Korbahti BK (2008) Photolytic Decolorization of Rose Bengal by UV/H2O2 and Data Optimization Using Response Surface Method. J of Hazard Mater 159:602–609

Robbins H, Monro S (1951) A Stochastic Approximation Method. Annals of Math Statistics 22:400–407

Sacks J, Welch WJ, Mitchell TJ, Wynn, HP (1989) Design and Analysis of Computer Experiments. Statistical Sci 4:409–435

Sereshti H, Karimi M, Samadi S (2009) Application of Response Surface Method for Optimization of Dispersive Liquid-liquid Microextraction of Water-soluble Components of Rosa Damascena Mill. Essential Oil. J of Chromatogr A 1216: 198–204

Sharma N (2008) Simulation Optimization Using Simulated Annealing: A Network-based Implementation and Study of Cooling Schedules. VDM Verlag

Simpson TW, Peplinski JD, Koch PN, Allen JK (2001) Metamodels for Computer-based Engineering Design: Survey and Recommendations. Eng with Comput 17:129–150

Sivanandam SN, Deepa SN (2007) Introduction to Genetic Algorithms. Springer

Sumathi S, Hamsapriya T, Surekha P (2008) Evolutionary Intelligence: An Introduction to Theory and Applications with Matlab. Springer

Tang Y, Zhou X, Chen J (2008) Preform Tool Shape Optimization and Redesign Based on Neural Network Response Surface Methodology. Finite Elem in Anal and Des 44:462–471

Tarley CRT, Silveira G, dos Santos WNL, Matos GD, da Silva EGP, Bezerra MA, Miro M, Ferreira SLC (2009) Chemometric Tools in Electroanalytical Chemistry: Methods for Optimization Based on Factorial Design and Response Surface Methodology. Microchem J 92:58–67

Tir M, Moulai-Mostefa N (2008) Optimization of Oil Removal from Oily Wastewater by Electrocoagulation Using Response Surface Method. J of Hazard Mater 158: 107–115

Varaajan S, Chen W, Pelka CJ (2000) Robust Concept Exploration of Propulsion Systems with Enhanced Model Approximation. Eng Optim 3:309–334

Vilakazi CB, Marwala T (2006) Bushing Fault Detection and Diagnosis Using Extension Neural Network. In: Proc of the 10th IEEE Intl Conf on Intell Eng Syst:170–174

Wang GG (2003) Adaptive Response Surface Method Using Inherited Latin Hypercube Design Points. J of Mech Des 125:210–220

Wei L, Yuying Y, Zhongwen X, Lihong Z (2009) Springback Control of Sheet Metal Forming Based on the Response Surface Method and Multi-objective Genetic Algorithm. Mater Sci and Eng A 499:325–328

Werbos PJ (1974) Beyond Regression: New Tool for Prediction and Analysis in the Behavioral Sciences. Unpublished Doctoral Dissertation, Harvard University, Cambridge

Yoon Y, Peterson LL (1990) Artificial Neural Networks: An Emerging New Technique. In: Proc of the 1990 ACM SIGBDP Conf on Trends and Directions in Expert Syst:417–422

Zeng G, Li SH, Yu ZQ, Lai XM (2009) Optimization Design of Roll Profiles for Cold Roll Forming

Chapter 7

Finite-element-model Updating Using a Hybrid Optimization Method

Abstract. This chapter presents a hybrid of particle-swarm optimization and the Nelder–Mead simplex optimization method for finite-element-model updating. It was observed, on average, that the hybrid of particle-swarm optimization and the Nelder–Mead simplex optimization method gives more accurate results, followed by the particle-swarm optimization and then the Nelder–Mead simplex method.

Keywords: Nelder–Mead, objective function, hybrid methods, particle-swarm optimization

7.1 Introduction

In previous chapters it was observed that the finite-element-model updating process is fundamentally, an optimization problem. The updating parameters are considered to be those parameters in the finite-element model that are thought to be very uncertain, while the objective function is a measure of the scalar distance between the finite-element predicted data and the measured data (Hua *et al.*, 2009). The measured data can use:

- the modal domain (Wang and Cai, 2009; Mthembu *et al.*, 2009);
- the frequency domain (Esfandiari *et al.* 2009);
- pseudo-modal energies (Marwala *et al.*, 2001);
- damping (Arora *et al.*, 2009); and
- the time–frequency domain (Marwala, 2002).

In the updating process, many parameters have been used, such as the modulus of elasticity, Poisson's ratios, and density (Adhikari and Friswell, 2010; Kozak *et al.*, 2009). This chapter uses the modulus of elasticity as the parameter to be updated.

The Nelder–Mead (NM) simplex method (Morgan *et al.*, 1990) and the Broyden–Fletcher–Goldfarb–Shanno (BFGS) method (Yuan *et al.*, 2009) were presented in Chapter 2 for finite-element-model updating. These methods are essentially local search methods and are good at identifying the local optimal solution. In Chapter 2, the NM simplex method was found to give more accurate results than the BFGS method when used for finite-element-model updating and applying these optimization methods to a simple beam and an unsymmetrical H-shaped structure.

There are many global optimization methods that can be used for the process of finite-element-model updating. One class of these is the class of evolutionary methods, inspired by natural processes. In previous chapters, two of these methods were considered being the genetic algorithm (GA) and particle-swarm optimization (PSO) (Meyer, 1996). PSO was inspired by the flocking of a population of birds, while GA was inspired by Darwin's theory of natural selection (Fogel, 1998). Another method, which is statistical in nature but inspired by the physical annealing process, is the simulated-annealing (SA) technique (Kirkpatrick *et al.*, 1983). In previous chapters, it was found that the PSO was better for finite-element-model updating than GA and SA when applied to the two examples considered in this book.

The response-surface method (RSM) is an approximate optimization technique (Marwala, 2004). In Chapter 6, it was found to provide similar results to those given by SA and the GA but with a much lower computational effort.

In this chapter a hybrid optimization method that combines PSO and the NM simplex method is presented and applied to finite-element-model updating. This hybrid method was implemented because it combines the advantages of a global search technique with those of a local search technique. Furthermore, the PSO method was used because of its successes in global searching, which was observed in the previous chapters, while the NM simplex method was also applied because of its demonstrated success in Chapter 2.

7.2 Introduction to Structural Dynamics

In this chapter, as in previous chapters, the moduli of elasticity of each element of the finite-element model were updated to ensure that the natural frequencies and mode shapes from the finite-element model reflect the measured data better. The natural frequencies and mode shapes were obtained from the measured vibration data through a process called modal analysis (Ewins, 2001).

Modal analysis is a procedure that is used to study the dynamic properties of structures using vibration data. It essentially entails measuring and analyzing the dynamic response of systems such as structures and fluids from excitation by an input force. For example, the dynamics of an airplane can be studied by measuring its vibration when it is excited by an electromagnetic shaker.

The analysis of the vibration data usually uses methods such as Fourier and wavelet analysis. From this analysis, data are extracted such as natural frequencies and mode shapes, which are used in this chapter.

In simple terms, the goal of the finite-element-model updating is to identify the moduli of elasticity that will give the more accurate updated finite-element model. The accuracy of the updated model is evaluated by the degree to which measurements reflect model predictions. To accomplish this aim, the following objective function to measure the distance between measured natural frequencies and the mode shapes and finite-element model predicted natural frequencies and mode shapes, was minimized:

$$E = \sum_{i=1}^{N} \gamma_i \left(\frac{\omega_i^m - \omega_i^{calc}}{\omega_i^m} \right)^2 + \beta \sum_{i=1}^{N} \left(1 - diag(MAC(\phi_i^{calc}, \phi_i^m))\right) \quad (7.1)$$

Here, m indicates a measured variable, calc indicates a calculated variable, ω_i is the ith natural frequency, ϕ_i is the ith mode shape vector, N is the number of modes and γ_i is the weighting factor that measures the relative distance between the initial estimated natural frequencies for mode i and the target frequency of the same mode. The parameter β is the weighting function imposed on the mode shapes.

As explained in Chapter 2, the MAC is the modal assurance criterion (Allemang and Brown, 1982) and diag(MAC)$_i$ stands for the ith diagonal element of the *MAC* matrix. The MAC is a measure of the correlation between two sets of mode shapes of the same dimension. In Equation 7.1, the first part serves the purpose of ensuring that the natural frequencies estimated by the finite-element model are as close to the measured ones as possible, while the second term ensures that the mode shapes between measurements and finite-element-model predictions are correlated.

7.3 Hybrid Particle-swarm Optimization and the Nelder–Mead Simplex

This chapter studies a hybrid of particle-swarm optimization (PSO) and the Nelder–Mead (NM) simplex method for finite-element-model updating. The hybrid procedure presented in this chapter is motivated by the need to create an optimization method that has two core characteristics. These are:

- the global search capability, which is normally found in evolutionary-type optimization such as genetic algorithms (GA) and PSO; and
- the local search capability that is intended to prevent premature convergence on a solution that is not necessarily optimal.

It was observed in Chapter 5 that of all the global optimization methods studied in this book, the PSO seems to give better results and it is relatively simple to implement. In Chapter 2 it was observed that of all the local optimization methods studied in this book, the NM simplex method seems to be the better method for local searching. Consequently, the hybrid method proposed in this chapter combines PSO with the NM simplex method.

As described in earlier chapters, the PSO is a stochastic, population-based optimization method inspired by the flocking of birds, schools of fish, and herds of animals. It is driven by individual and group intelligence. These swarms preserve an "evolutionary advantage" by either adapting to their environment, avoiding predators or finding rich sources of food, all by the process of "information sharing" (Marwala, 2009). The PSO is initialized by generating a random population of possible solutions (particles). Each solution's fitness is evaluated, according to an objective function in Equation 7.1, and the fittest or best solution is stored. All the solutions in the problem space have their own set of coordinates and these coordinates are linked with the fittest solution encountered thus far. Another value stored is the best solution encountered in the neighborhood of the particular solutions or particles. The PSO technique entails accelerating the velocities of each particle in the direction of the best solutions the particles have found so far and the best solutions the swarm has found.

Flores *et al.* (2004) applied the PSO for force identification in mechanical systems. Other successful applications of PSO include studies in:

- hydroelectric system scheduling (He *et al.*, 2009);
- production planning (Chen and Lin, 2009);
- the prediction of silicon content in hot metal (Tang *et al.*, 2009);
- geomechanical systems (Zhao and Yin, 2009);
- electric power system (AlRashidi and El-Hawary, 2009);
- ice-storage air-conditioning (Lee *et al.*, 2009);
- in control systems (dos Santos Coelho and Coelho, 2009); and
- heat exchangers (Silva *et al.*, 2009).

The NM simplex method, originally introduced by Nelder and Mead (1965), is a derivative-free algorithm that is used for multi-dimensional unconstrained problems. This method uses simplex vertices. The simplex vertices coordinates are altered by using reflection, expansion and contraction operators. The basic method is simple to implement and understand, so for this reason, it has been used widely. Because it is derivative free, it is also suitable for nonsmooth functions. The NM simplex method has been used in:

- parameter estimation where the function values are uncertain and noisy (Camp and Garbe, 2004);
- discontinuous functions (Chen *et al.*, 2009);
- winter-wheat irrigation (Shang and Mao, 2006);
- chemical engineering (Wu and Ierapetritou, 2006); and
- buffer placement in serial lines (Harris and Powell, 1999).

Hybrid optimization methods have not been widely used for finite-element-model updating. Kang *et al.* (2009) applied a hybrid of the NM simplex method and artificial bee-colony optimization for parameter identification in concrete dam-foundation systems. The results obtained demonstrated that the proposed method performed better than a GA.

Lin and Hsieh (2009) introduced an efficient hybrid method that combines Taguchi, PSO and GA and applied this to the simulation of protein folding. The PSO and GA were used because of their ability to identify global optimal solutions, whereas the Taguchi technique was used because of its ability to use the optimum offspring. Additionally, the PSO method, which was motivated by a mutation method in GA, was also implemented. The results obtained showed that this method gives better results than the current evolutionary methods.

Kumar and Balasubramanian (2009) applied a hybrid PSO for kinetic parameter estimation in hydrocracking. Niknam and Firouzi (2009) applied a combination of PSO and the NM simplex method for state estimation in renewable-energy sources. This hybrid method was found to successfully approximate load and renewable energy sources. When this method was compared to the standard PSO, honeybee-mating optimization (HBMO), neural networks, ant-colony optimization (ACO), and a GA it was found to be very useful and efficient for complex problems.

Chen *et al.* (2010) proposed a novel hybrid particle-swarm optimizer with an external optimization technique. This hybrid procedure was proposed because the PSO method was found to have untimely convergence, in particular in complex multi-modal functions. External optimization is a successful optimization method that has been proposed in recent times and is a local-search heuristic technique. The proposed hybrid method unites the exploration capacity of the PSO method with the exploitation capability of extremal optimization. The hybrid method was found to give better results than these methods used in isolation. This is because it prevents premature convergence.

Goswami and Acharjee (2010) applied a hybrid PSO technique for multiple low-voltage power-flow solutions. This procedure entails using the PSO method for finding the initial estimation of the rectangular Newton–Raphson power flow and then using the rectangular Newton–Raphson load flow to calculate a low-voltage solution. Thereafter, optimal multipliers were used to estimate further low-voltage solutions.

Begambre and Laier (2009) presented a hybrid of the PSO and the NM simplex method for structural-damage identification. This method was used to minimize the objective function that used frequency-response functions. The hybrid method directed the PSO parameters using the NM simplex method. This ensured that the convergence of the PSO method was not dependent on the heuristic constants and, consequently the stability and confidence of the method was improved. When this hybrid method was implemented, it was found to perform better than the simulated annealing (SA) and PSO. This procedure was successfully applied in a damage-detection procedure and in a nonlinear oscillator.

Yang *et al.* (2009) applied a hybrid data clustering technique based on K-harmonic means and PSO. The proposed method ensured that the clustering avoided local optima and early convergence. The hybrid method was found to perform better than the methods used in isolation.

Yu *et al.* (2009) applied hybrid PSO and sequential number-theoretic optimization method for localization of acoustic sources in sensor networks. The results found showed that the hybrid method achieved strong convergence with complicated inference performance.

Niknam (2009) implemented an efficient hybrid of HBMO and the discrete particle-swarm optimization (DPSO) for multi-objective distribution feeder reconfiguration, which has a multi-objective and nondifferentiable objective function. The results from the presented hybrid method were better than when the DPSO and HBMO were used in isolation.

Marinakis and Marinaki (2010) implemented a hybrid of the GA and PSO for vehicle-routing problems. In this hybrid method, the PSO was used for the evolution of each individual of the total population, which included the parents and the offspring, to improve each particle's physical move to attain the prerequisites to be selected as a parent. This increases the probability that the information from each parent was conveyed to its offspring and that of the whole population. This hybrid method could thus explore the solution space more effectively than the traditional PSO and GA used in isolation.

Liao *et al.* (2009) applied a hybrid of the PSO and fuzzy neural networks to control the temperature of a reheating furnace. The structure of the controller consisted of a fuzzy c-means clustering method. A hybrid PSO was used to optimize the weights. The hybrid technique combined the global characteristics of the density-based selection approach and the exact exploration of a clonal expansion in an immune system together with a rapid local exploration of the PSO. The hybrid method was successfully used to minimize fuel consumption, the temperature gradient inside a billet, and the error between the mean and target temperatures of a billet at the furnace outlet.

Mahmood (2009) used a hybrid of PSO for locating duplicates of articles in a dispersed web-server system. The hybrid PSO method that was used exploited the global search capability of the PSO and the local search capacity of Tabu search to attain solutions of high quality. The hybrid method was found to outperform the GA, PSO, and Tabu search.

Wei and Xingsheng (2009) introduced a hybrid PSO method where the average position and velocity of particles were included into the PSO method together with a differential evolution computation, which brings in an additional population for prior crossover. The prior crossover was intended to suitably diversify the population and thereby boost the probability that a global optimal solution was reached. A differential evolution section considered the stochastic differential variation, and improved the exploitation in the neighborhoods of the current solution.

Zahara and Kao (2009) applied a hybrid of PSO and the NM simplex method for constrained engineering design problems. The results revealed that the hybrid method was very efficient and successful in identifying optimal solutions.

Sadati *et al.* (2009) applied a hybrid of the PSO and SA for under-voltage load-shedding problems, while Cui (2008) applied a hybrid of the SA and PSO for power coal-blending optimization.

Chen *et al.* (2008) implemented a novel hybrid of PSO and an artificial fish-swarm algorithm for identifying the network weights in a neural network. The hybrid model was found to be more accurate and stable than the Levenberg–Marquardt optimization.

Zhao *et al.* (2008) applied a hybrid of an improved PSO, Powell and pattern search in a parallel fashion and found that the improved PSO technique gave more stability and a faster convergence than the traditional PSO method, while the hybrid PSO was found to solve complex global optimization problems and conduct parallel operations better. The hybrid PSO is illustrated in Figure 7.1.

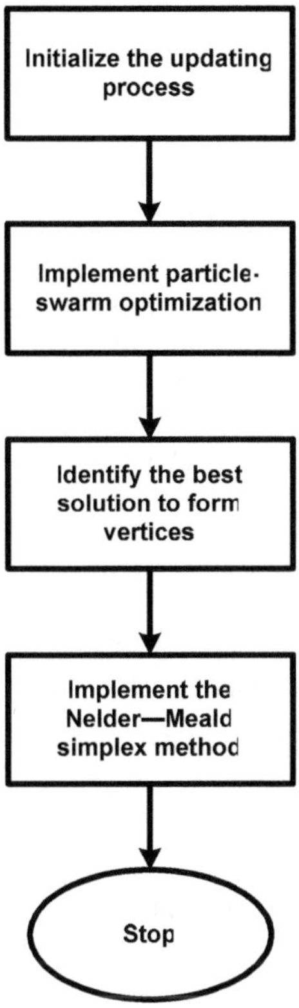

Figure 7.1 The hybrid of the PSO and the NM simplex method

Other successful implementations of the hybrid procedures making use of PSO include a hybrid of:

- PSO and a discrete position update method, which was applied on manufacturing cell design (Duran *et al.*, 2008);
- PSO, which uses the behavior of insect swarms and the natural-selection mechanism and was applied to kinematics control (Wen *et al.*, 2008);
- the heuristic route exploration algorithm and PSO for the logistics network design (Qin and Qin, 2008);
- the PSO and GA mutation applied to spatial clustering with obstacles constraints (Zhang *et al.*, 2008a);
- an improved ant-colony optimization (ACO) and PSO for spatial data clustering (Zhang *et al.*, 2008b); and

- a PSO that incorporates the principle of the thermodynamic theory, rating-based entropy and a component thermodynamic replacement (Wu *et al.*, 2008).

This chapter implements a hybrid of the PSO method. The advantages of this hybrid method are that:

- it can identify a global optimum solution;
- it avoids premature convergence;
- it has increased convergence accuracy; and
- it improves convergence speed.

The disadvantage of this hybrid approach is that it involves the arbitrary swarm-optimization parameter-selection problem.

This hybrid of PSO and the NM simplex method is conceptually illustrated in Figure 7.1, which is explained as follows:

1. Initialize a population of particles' positions and velocities. The positions of the particles must be randomly distributed in the updating parameter space.
2. Calculate the velocity for each particle in the swarm using: $v_i(k+1) = \gamma v_i(k) + c_1 r_1 (pbest_i - p_i(k)) + c_2 r_2 (gbest(k) - p_i(k))$ where $v_i(k)$ is the velocity at Step k, $pbest_i$ is the best particle encountered by the particle, $gbest(k)$ is the best particle position identified in the swarm, γ is the inertia of the particle, c_1 and c_2 are the "trust" parameters, and r_1 and r_2 are random numbers between 0 and 1.
3. Update the position p_i of the ith particle using $p_i(k+1) = p_i(k) + v_i(k+1)$.
4. Repeat Steps 2 and 3 until convergence.
5. Pick the three best results from the swarm that were identified in Step 3, and use this to form a simplex vertex.
6. For every iteration in the optimization procedure, replace the vertex with the worst fitness measure as defined by the objective function in Equation 7.1 with a new vertex.
7. The coordinates of the new vertex are established by reflecting the old vertex's point about the outstanding vertices.
8. If the fitness measure of the current vertices is lower than the preceding removed vertex's fitness, the dimensions of the simplex are minimized. If not, they are enlarged.
9. The process in Steps 6 to 8 is continued until the vertices functional evaluation values become similar.

7.4 Example 1: Simple Beam

The aluminum beam shown in Figure 2.2, was used to test the hybrid of PSO and the NM simplex method for finite-element-model updating. The details of this beam were reported by Marwala (1997). The beam was tested freely suspended using elastic rubber bands. The beam was excited using an electromagnetic shaker and the response was measured using an accelerometer. The beam was also modeled using the structural dynamics toolbox (Balmès, 1997) and the beam was divided into 11 elements. The finite-element-model used Euler–Bernoulli beam elements. The moduli of elasticity of these elements were used as updating parameters. When the finite-element-model updating was implemented the moduli of elasticity was restricted to vary from 6×10^{10} to 8×10^{10} N m^{-2}. It was excited at various positions, and the acceleration was measured at 10 positions. A set of 10 frequency-response functions were calculated and a roving accelerometer was used for testing. The moduli of elasticity of these elements were used as updating parameters. When the finite-element-model updating was implemented, the moduli of elasticity were restricted to vary from 6×10^{10} to 8×10^{10} N m^{-2}. The weighting factor in the first term of Equation 7.1 was calculated for each mode as the square of the error between the measured natural frequency and the natural frequency calculated from the initial model. The weighting function for the second term in Equation 7.1 was set to 0.75. The hybrid of PSO and the NM simplex optimization method, the PSO (in isolation) and the NM simplex method (in isolation) were implemented to optimize Equation 7.1 and the results in Table 7.1 were obtained.

On implementing the PSO for both the hybrid method and in isolation for finite-element-model updating the following parameters were chosen:

Population size = 50, generations = 200, $c_1 = 0.05$, $c_2 = 0.01$, and $w = 0.002$.

Table 7.1 Results showing measured frequencies of the beam, the initial frequencies and the frequencies obtained when the finite-element model of a beam is updated using a hybrid of the PSO and the NM simplex method, NM (in isolation) and PSO (in isolation)

Modes	Measured frequency (Hz)	Initial frequency (Hz)	Frequencies from the NM updated model (Hz)	Frequencies from the PSO updated model (Hz)	Frequencies from the hybrid updated model (Hz)
1	41.5	42.3	41.7	41.5	41.9
2	114.5	117.0	114.7	112.4	114.3
3	224.5	227.3	221.3	224.4	225.3
4	371.6	376.9	370.3	370.9	372.3

When a modulus of elasticity of 7×10^{10} N m^{-2} was assumed, the error between the first measured natural frequency and that from the initial finite-element model was 1.9%. When the PSO was used for finite-element-model updating, this error was reduced to 0.0%, on using the NM simplex method it was reduced to 0.5% and when the hybrid was used, it was reduced to 1.0%.

The error between the second measured natural frequency and that from the initial model was 2.2%. When the PSO was used, this error was reduced to 1.8%. When using the NM simplex method it was reduced to 0.2%, and when using a hybrid approach it was reduced to 0.2%.

The error of the third natural frequencies between the measured data and the initial finite-element model was 1.2%. When the PSO method was used, this error was reduced to 0.0% and using the NM simplex method, it was increased to 1.4%, while the hybrid method reduced it to 0.4%.

The error between the fourth measured natural frequency and that from the initial model was 1.4%. When the PSO technique was used, this error was reduced to 0.2% using the NM simplex method it was reduced to 0.3% and using the hybrid approach reduced it to 0.2%.

Overall, the hybrid method gave the best results followed by the PSO method. On average, all three methods were improvements when compared to the average error between the initial finite-element model and the measured data calculated over four modes of 1.7%.

The updated models implemented were also validated on the mode shapes they predicted. To make this assessment possible, the modal assurance criterion (MAC) was used (Allemang and Brown, 1982). The mean of the diagonal of the MAC vector was used to compare the mode shapes predicted by the updated and initial finite-element models with the measured mode shapes. The average MAC calculated between the mode shapes from an initial finite-element model and the measured mode shapes was 0.9986. When the average MAC was calculated between the measured data and data obtained from the updated finite-element models, the results in Table 7.2 were obtained. It was observed that the hybrid updated finite-element model gave the best-improved averages of the diagonals of the MAC matrix followed by PSO method.

Table 7.2 Beam results showing the MAC calculated between measured mode shapes and the initial finite-element model, the hybrid of PSO and the NM simplex, NM (in isolation) and PSO (in isolation)

Method	MAC
Initial model	0.9986
NM	0.9988
PSO	0.9989
Hybrid method	0.9990

7.5 Example 2: Unsymmetrical H-shaped Structure

An unsymmetrical H-shaped aluminum structure described in Chapter 2 was also used to validate the proposed method (Marwala, 1997). This structure was used and the initial finite-element model gave data that were far from the measured data. The structure was suspended using elastic rubber bands, excited using an electromagnetic shaker and the response was measured using an accelerometer. The structure was divided into 12 elements. It was excited and acceleration was measured at 15 positions indicated by the arrows. The structure was tested freely suspended, and a set of 15 frequency-response functions were calculated. A roving accelerometer was used for the testing. The mass of the accelerometer was found to be negligible compared to the mass of the structure. As in the previous example, the finite-element model was constructed with the structural dynamics toolbox (Balmès, 1997) using the Euler–Bernoulli beam elements (Zienkiewicz, 1971).

The finite-element model contained 12 elements. As in the previous example, the moduli of elasticity of these elements were used as updating parameters that were restricted to fall in the interval 6×10^{10} to 8×10^{10} N m^{-2}. The weighting factor, in the first term of Equation 4.2 was calculated for each mode as the square of the error between the measured natural frequency and the natural frequency calculated from the initial model. The weighting function for the second term in Equation 7.1 was set to 0.75. A hybrid of the PSO and the NM simplex optimization method, PSO (in isolation) and the NM simplex method (in isolation) were implemented to optimize Equation 7.1.

When these optimization methods were implemented for finite-element-model updating, the results shown in Tables 7.3 and 7.4 were obtained. Table 7.3 shows the measured natural frequencies, initial natural frequencies and natural frequencies obtained by a hybrid of the PSO and the NM simplex optimization method, the PSO (in isolation) and the NM simplex method (in isolation) that were implemented to update the finite-element models.

When a modulus of elasticity of 7×10^{10} N m^{-2} was assumed, the error between the first measured natural frequency and that from the initial finite-element model was 4.3%. When the PSO was used for finite-element-model updating, this error was reduced to 0%; the NM simplex method reduced this error to 3.3% and the hybrid method reduced it to 0.6%.

The error between the second measured natural frequency and that from the initial model was 8.4%. When the PSO was used, this error was reduced to 0.4%; the NM simplex method reduced this error to 1.8% and the hybrid method reduced it to 0.7%.

The error in the third natural frequencies between the measured data and the initial finite-element model was 9.6%. When the PSO method was used, this error was reduced to 0.1%; using the NM simplex method it was reduced to 1.9%, while using the hybrid technique reduced it to 0.0%.

The error between the fourth measured natural frequency and that from the initial model was 3.7%. When the PSO was used, this error was reduced to zero; using the NM simplex method it was reduced to 1.0%, while the hybrid method reduced it to 0.0%.

The error between the fifth measured natural frequency and that from the initial model was 1.6%. When the PSO was used, this error was reduced to 1.5%; the NM simplex method increased it to 2.6%, while the hybrid technique reduced it to 0.3%.

Overall, the hybrid method gave the best results with an average error, calculated over all the five natural frequencies of 0.3%. The PSO gave the results with an average error of 0.4%, while the NM simplex method gave an average error of 2.1%. On average, all three methods were improvements on the average error between the initial finite element model and the natural frequencies, which was 5.5%.

Table 7.3 Results showing measured frequencies of the beam, the initial frequencies and the frequencies obtained when the finite-element model of a beam was updated using a hybrid of the PSO and the NM simplex method, the NM (in isolation) and the PSO (in isolation)

Modes	Measured frequency (Hz)	Initial frequency (Hz)	Frequency from the NM updated model (Hz)	Frequency from the PSO updated model (Hz)	Frequency from a hybrid updated model (Hz)
1	53.9	56.2	52.1	53.9	53.6
2	117.3	127.1	119.4	117.8	118.1
3	208.4	228.4	212.4	208.5	208.5
4	254.0	263.4	251.3	253.9	254.1
5	445.1	452.4	433.6	438.5	443.7

As in the previous example, the updated models implemented were validated on the mode shapes they predicted using the MAC. The hybrid of PSO and the NM simplex optimization method, PSO (in isolation) and the NM simplex method (in isolation) for updated finite-element models gave improved averages for the diagonals of the MAC matrices with the hybrid method showing the best results, followed by the PSO method.

Table 7.4 Beam results showing the MAC calculated between the measured mode shapes and the initial finite-element model, the hybrid PSO and the NM simplex method, the NM (in isolation) and the PSO (in isolation)

Method	MAC
Initial model	0.8394
NM	0.8404
PSO	0.8434
Hybrid method	0.8440

7.6 Conclusion

In this chapter, a hybrid of the PSO and the NM simplex optimization method, the PSO (in isolation) and the NM simplex method (in isolation) were implemented for finite-element-model updating. When these techniques were tested on a simple beam and an unsymmetrical H-shaped structure, it was observed, on average, that the hybrid of the PSO and the NM simplex optimization method gave results that were more accurate. This was followed by the PSO and then the NM simplex method.

7.7 Future Work

This chapter introduced a hybrid of the PSO and the NM simplex optimization methods for finite-element updating. This hybrid was a simple combination of the two methods where the PSO identify an optimal updated model that is refined using the NM. For future work, other hybrid procedures should be explored.

References

Adhikari S, Friswell MI (2010) Distributed Parameter Model Updating Using the Karhunen–Loève Expansion. Mech Syst and Signal Process, 24:2. 326-339

Allemang RJ, Brown DL (1982) A Correlation Coefficient for Modal Vector Analysis. In: Proc of the 1st Intl Modal Analysis Conf 1–18

AlRashidi MR, El-Hawary ME (2009) A Survey of Particle Swarm Optimization Applications in Electric Power Systems. IEEE Trans on Evol Comput 13:913–918

Arora V., Singh SP, Kundra TK (2009) Finite Element Model Updating with Damping Identification. Journal of Sound and Vibration, 324: 111–1123

Balmès E (1997) Structural Dynamics Toolbox User's Manual version 2.1. Sèvres, Scientific Software Group, France

Begambre O, Laier JE (2009) A Hybrid Particle Swarm Optimization – Simplex Algorithm (PSOS) for Structural Damage Identification. Adv in Eng Softw 40:883–891

Camp M, Garbe H (2004) Parameter Estimation of Double Exponential Pulses (EMP, UWB) With Least Squares and Nelder Mead Algorithm. IEEE Trans on Electromagn Compat 46:675–678

Chen HK, Hsueh CF, Chang MS (2009) Production Scheduling and Vehicle Routing with Time Windows for Perishable Food Products. Comput and Oper Res 36:2311–2319

Chen MR, Li X, Zhang X, Lu YZ (2010) A Novel Particle Swarm Optimizer Hybridized with Extremal Optimization. Appl Softw Comput J 10:367–373

Chen YY, Lin JT (2009) A Modified Particle Swarm Optimization for Production Planning Problems in the TFT Array Process. Expert Syst with Appl 36:12264–12271

Chen X, Sun D, Wang J, Liang J (2008) A Novel Hybrid Evolutionary Algorithm Based on PSO and AFSA for Feedforward Neural Network Training. Intl Conf on Wireless Commun, Netw and Mob Comput:Paper ID - 4680707

Cui Y (2008) Application of Simulated Annealing Particle Swarm Optimization Algorithm in Power Coal Blending Optimization. Intl Conf on Wireless Commun, Netw and Mob Comput:Paper ID - 4679158

dos Santos Coelho L, Coelho AAR (2009) Model-free Adaptive Control Optimization Using a Chaotic Particle Swarm Approach. Chaos, Solitons and Fractals 41:2001–9

Duran O, Rodriguez N, Consalter LA (2008) Hybridization of PSO and a Discrete Position Update Scheme Techniques (stet) for Manufacturing Cell Design. Lecture Notes in Computer Science (including subseries of Lect Notes in Artif Intell and Lect Notes in Bioinformatics) 5317:503–512

Esfandiari AF, Bakhtiari-Nejad, Rahai A, Sanayei A (2009) Structural Model Updating Using Frequency Response Function and Quasi-Linear Sensitivity Equation. J of Sound and Vib 326:557–573

Ewins DJ (2001) Modal Testing. Theory, Practice and Application (Mechanical Engineering Research Studies: Engineering Dynamics Series) Wiley

Flores JER, Viana FAC, Rade DA, Steffen Jr V (2004) Force Identification of Mechanical Systems by Using Particle Swarm Optimization. 10th AIAA/ISSMO Multidiscip Analysis and Optim Conf 5:3105–3116

Fogel DB (1998) Evolutionary Computation. The Foss Rec, IEEE Press, N.Y

Goswami SK, Acharjee P (2010) Multiple Low Voltage Power Flow Solutions Using Hybrid PSO and Optimal Multiplier Method. Expert Syst with Appl 37:2473–2476

Harris JH, Powell SG (1999) An Algorithm for Optimal Buffer Placement in Reliable Serial Lines. IIE Trans,Inst of Ind Eng 31:287–302

He YY, Zhou JZ, Xiang XQ, Chen H, Qin H (2009) Comparison of Different Chaotic Maps in Particle Swarm Optimization Algorithm for long-term Cascaded Hydroelectric System Scheduling. Chaos, Solitons and Fractals 42:3169–3176

Hua XG, Ni YQ, Kon JM (2009) Adaptive Regularization Parameter Optimization in Output-Error-Based Finite Element Model Updating. Mech Syst and Signal Process 23:563–579

Kang F, Li J, Xu Q (2009) Structural Inverse Analysis by Hybrid Simplex Artificial Bee Colony Algorithms. Comput and Struct 87:861–870

Kirkpatrick S, Gelatt CD, Vecchi MP (1983) Optimization by Simulated Annealing. Sci, New Series 220: 671–680

Kozak MT, Öztürk M, Özgüven HN (2009) A Method in Model Updating using Miscorrelation Index Sensitivity (2009) Mech Syst and Signal Process, 23: 1747–1758

Kumar V, Balasubramanian P (2009) Kinetic Parameter Estimation in Hydrocracking Using Hybrid Particle Swarm Optimization. Fuel 88:2171–2180

Lee WS, Chen Y, Wu TH (2009) Optimization for Ice-storage Air-conditioning System Using Particle Swarm Algorithm. Appl Energy 86:1589–1595

Liao YX, She JH, Wu M (2009) Integrated Hybrid-PSO and Fuzzy-NN Decoupling Control for Temperature of Reheating Furnace. IEEE Trans on Ind Electron 56:2704–2714

Lin CJ, Hsieh MH (2009) An Efficient Hybrid Taguchi-genetic Algorithm for Protein Folding Simulation. Expert Syst with Appl 36:12446–12453

Mahmood A (2009) Replicating Web Contents Using a Hybrid Particle Swarm. Optim Inf Process and Manag, in press

Marinakis Y, Marinaki M (2010) A Hybrid Genetic–Particle Swarm Optimization Algorithm for the Vehicle Routing Problem. Expert Syst with Appl 37:1445–1455

Marwala T (1997) A Multiple Criterion Updating Method for Damage Detection on Structures. Master's Thesis, University of Pretoria

Marwala T (2001) On Fault Identification Using Pseudo-modal Energies and Modal Properties. Am Inst of Aeronaut and Astronaut J 39:1608–1617

Marwala T (2002) Finite Element Updating Using Wavelet Data and Genetic Algorithm. Am Inst of Aeronaut and Astronaut, J of Aircr 39:709–711

Marwala T (2004) Finite Element Model Updating Using Response Surface Method. In: Proc of the 45th AIAA/ASME/ASCE/AHS/ASC Struct, Struct Dyn & Mater Conf USA

Marwala T (2009) Computational Intelligence for Missing Data Imputation, Estimation and Management: Knowledge Optimization Techniques. Information Science Reference Imprint, IGI Global Publications, New York

Marwala T, Adhikari S, Heyns PS (2001) Dynamic Model Updating Using Pseudo Modal Energies. In: Proc of the 19th Intl Modal Analyis Conf, Kissimmcc

Meyer J-A (1996) Artificial Life and the Animal Approach to Artificial Intelligence. Artificial Intelligence, 325–354

Morgan E, Burton KW, Nickless G (1990) Optimization Using the Modified Simplex Method. Chemometrics and Intelligent Laboratory Systems, 7:209–222

Mthembu L Marwala T, Friswell MI, Adhikari S (2009) Bayesian Evidence for Finite Element Model Updating. In: Proc of the IMAC XXVII, Florida

Nelder JA, Mead R (1965) A Simplex Method for Function Minimization. Comput J 7:308–313

Niknam T (2009) An Efficient Hybrid Evolutionary Algorithm Based on PSO and HBMO Algorithms for Multi-objective Distribution Feeder Reconfiguration. Energy Convers and Manag 50:2074–2082

Niknam T, Firouzi BB (2009) A Practical Algorithm for Distribution State Estimation Including Renewable Energy Sources. Renew Energy 34:2309–2316

Qin XW, Qin F (2008) Hybrid Particle Swarm Optimization Algorithm for the Logistics Network Design Problem Under Concave Cost Function. In: Proc of the Intl Conf on Comput Sci and Softw Eng 4721964:1182–1186

Sadati N, Amraee T, Ranjbar AM (2009) A Global Particle Swarm-Based-Simulated Annealing Optimization Technique for Under-voltage Load Shedding Problem. Appl Softw Comput J 9:652–657

Shang S, Mao X (2006) Application of a Simulation Based Optimization Model for Winter Wheat Irrigation Scheduling in North China. Agric Water Manag 85:314–322

Silva AP, Ravagnani MASS, Biscaia Jr EC, Caballero, JA (2009) Optimal Heat Exchanger Network Synthesis Using Particle Swarm Optimization. Optim and Eng, in press

Tang X, Zhuang L, Jiang C (2009) Prediction of Silicon Content in Hot Metal Using Support Vector Regression Based on Chaos Particle Swarm Optimization. Expert Syst with Appl 36:11853–11857

Wang Y, Cai Z (2009) A Hybrid Multi-swarm Particle Swarm Optimization to Solve Constrained Optimization Problems. Front of Comput Sci in China 3:38–52

Wei X, Xingsheng G (2009) A Hybrid Particle Swarm Optimization Approach with Prior Crossover Differential Evolution. In: Proc of the 1st ACM / SIGEVO Summit on Genet and Evol Comput :671–677

Wen X, Sheng D, Huang J (2008) A Hybrid Particle Swarm Optimization for Manipulator Inverse Kinematics Control. Lect Notes in Comput Sci (Including Subseries Lect Notes in Artif Intell and Lect Notes in Bioinformatics) 5226:784–791

Wu D, Ierapetritou M (2006) Lagrangean Decomposition Using an Improved Nelder-Mead Approach for Lagrangean Multiplier Update. Comput and Chem Eng 30:778–789

Wu Y, Li Y, Xu X, Peng S (2008) Hybrid Particle Swarm Optimization Based on Thermodynamic Mechanism. Lect Notes in Comput Sci (Including Subser Lect Notes in Artif Intell and Lect Notes in Bioinformatics) 5361:279–288

Yang F, Sun T, Zhang C (2009) An Efficient Hybrid Data Clustering Method Based on K-harmonic Means and Particle Swarm Optimization. Expert Syst with Appl 36:9847–9852

Yu Z, Wei J, Liu H (2009) Force-directed Hybrid PSO-SNTO Algorithm for Acoustic Source Localization in Sensor Networks. Signal Process 89:1671–1677

Yuan Y, Dai H (2009) The Direct Updating of Damping and Gyroscopic Matrices. J of Comput and Appl Math 231:255–261

Zahara E, Kao YT (2009) Hybrid Nelder-Mead Simplex Search and Particle Swarm Optimization for Constrained Engineering Design Problems. Expert Syst with Appl 36:3880–3886

Zhang W, Yu W, Yang Z (2008a) Genetic and Particle Swarm Algorithm-Based Optimization Solution for High-Dimension Complex Functions. In: Proc of the 4th Intl Conf on Nat Comput: 511–515

Zhang X, Liu Y, Wang J, Deng G, Zhang C (2008b) Hybrid Particle Swarm Optimization with GA Mutation to Solve Spatial Clustering with Obstacles Constraints. In Proc of the Intl Symp on Comput Intell and Des: 299–302

Zhao HB, Yin S (2009) Geo-Mechanical Parameters Identification by Particle Swarm Optimization and Support Vector Machine. Appl Math Model 33:3997–4012

Zhao Y, An X, Luo W (2008) Hybrid Particle Swarm Optimization Based on Parallel Collaboration.In: Procof the Intl Conf on Intell Comput Technol and Autom 1:65–70

Zienkiewicz OC (1971) The Finite Element Method in Engineering Science. McGraw-Hill, London

Chapter 8

Finite-element-model Updating Using a Multi-criteria Method

Abstract. In this chapter a multiple criteria method (MCM) is presented and tested for finite-element-model updating using a simple beam with holes and an irregular H-shaped structure. The MCM minimizes the Euclidean norm of the error matrix that combines the modal property data with the frequency-response function data. The results given by the MCM are compared with the results from a frequency-response function method (FRFM), and those obtained from the modal property method (MPM). The three methods were compared based on their ability to reproduce the measured parameters. It was observed, on average, that the MCM gives the best results.

Keywords: multi-criteria method, modal property method, frequency-response method, modal properties, multi-criteria optimization

8.1 Introduction

In this chapter, a multiple criteria method (MCM) is used to update a finite-element model for it to reflect the measured data better (Marwala and Heyns, 1998). To fully understand how updating is performed, it is appropriate to first discuss the theoretical foundation of modal analysis.

Any given vibrating structure may be described in terms of three model types: a *spatial model*, a *modal model*, and a *response model*. Spatial, modal, and response models, respectively, entail the expression of a dynamic system in terms of (Ewins, 1986):

- mass, damping, and stiffness matrices;
- natural frequencies and mode shapes; and
- frequency or impulse response functions.

The experimental route to vibration analysis is known as *experimental modal analysis*. The spatial model is usually constructed from the finite-element model, while the response model is obtained from an experiment.

Frequency-response functions (FRFs) are measured by artificially exciting the structure and measuring the corresponding responses. The set of FRF measurements may then be used to determine the mode shapes at different resonant frequencies. Many methods have been successfully used to extract modal properties from measured FRFs.

Measured FRFs and mode shapes have been intensively used in finite-element-model updating, by exploiting their characteristics. FRFs are measured directly and, therefore, considerable labor is saved by using them directly for finite-element-model updating. Alternatively, the problem of updating may be approached by using the measured mode shapes and natural frequencies. Even though both methods can be used to give an updated finite-element model, it often occurs that the solutions that are given by the modal approach are different from the solution given by the FRFs approach. This is because the updating techniques do not give a unique finite-element model. In this chapter it was assumed that solving the updating problem from two different perspectives – the measured FRFs approach and the measured modal properties approach – would increase the chances of obtaining a representative finite-element-model updating solution.

The measured FRFs were substituted into the equation of motion obtained from the finite-element model, and the resulting error was measured. The Euclidean norm (Burden and Faires, 1993) of the error was minimized by employing an optimization procedure and by engaging the physical parameters that were not accurately known as design variables. In the second phase of this work, the measured mode shapes and natural frequencies were substituted into the equation derived from the orthogonality properties, where the mass and stiffness matrices from the finite-element model are used. The Euclidean norm error due to the combination of measured mode shapes, and natural frequencies with finite-element mass and stiffness matrices were minimized by taking inaccurate physical parameters as design variables. In the third phase of this study, the Euclidean norms of errors obtained from the FRF approach and the orthogonality approach were added using equal weighting functions and the resulting error was minimized by taking physical parameters as design variables.

The FRF approach was based on the work done by D'Ambrogio and Zobel, 1990; while the modal properties approach was based on the modified version of the orthogonality properties.

8.2 Mathematical Foundation

Vibration data can be represented in many domains. These include time, frequency, modal and time–frequency domains. For example, Marwala (2002) successfully used wavelet data, which is in the time–frequency domain, for finite-element updating. Hemez and Doebling (1999) successfully used time-domain data for finite-element-model updating. In this chapter, a frequency-domain method and a modal-domain method were used individually and collectively for finite-element model updating. Other collective use of different domain data includes time- and modal-domain methods that were used for finite-element-model updating by Link and Weiland (2008), who applied this to damage identification.

8.2.1 Frequency-response Function Method (FRFM)

The frequency-response function method (FRFM) directly uses the measured FRFs for finite-element-model updating. Using the FRFs directly has the following advantages (Dascotte and Strobbe, 1999; Ren and Beards, 1995):

- it eliminates the error that is usually incurred during the extraction of modal properties from measured FRFs;
- in the absence of noise, the FRFs represent the equation of motion at each frequency point; and
- the FRFs contain damping properties that otherwise have to be modeled when using the modal property data.

D'Ambrogio and Zobel (1990) used the measured FRFs directly to update the finite-element model by updating the stiffness matrices. Esfandiari *et al.* (2009) used the FRF's data for model updating by estimating the mass and stiffness matrices of a finite-element model as well as the damping characteristics. A least-squares process with a suitable normalization procedure was implemented to solve the over-determined system of equations with noisy vibration data. The sensitivity equation and appropriate choice of measured frequency data enhanced the precision and convergence of the finite-element-model updating process.

Arora *et al.* (2009) used the FRF's data to update a finite-element model and the damping identification. The updated finite-element model could predict the measured FRFs.

Sadr *et al.* (2007) used the FRF data and a neural-network method for finite-element-model updating. The input to the neural network was the FRF data, while the output was the updating parameters. The disadvantage of many frequency points was dealt with by using a principal component analysis.

Lin and Zhu (2007) applied response-function data under base excitation to identify mass and stiffness modelling, and thus updated the finite-element model. This method was resistant to noise.

Ziaei-Rad (2005) applied the FRF data, genetic algorithm (GA), and adaptive simulated-annealing (SA) optimization methods for updating the finite-element model of rotating structures. A regular linear least-squares formulation was implemented to update the element mass, damping, gyroscopic, and stiffness matrices using FRFs data from a particular multiplier and the updating parameters.

Asma and Bouazzouni (2005) proposed and successfully applied an updating technique based on measured FRFs where the objective function was the difference between the measured and the analytical frequency responses. The updating parameters were the coefficients that were related to the elements of the mass and stiffness matrices.

Tong *et al.* (2005) proposed a finite-element-updating method that is based on the FRFs and applied this to submarines. Other applications of the FRFs to finite-element-model updating include:

- Lu and Tu (2004) applied FRFs and neural networks;
- Nobari and Farjoo (2004) used FRF methods;
- Kerschen and Golinval (2004) applied FRFs to deal with nonlinearities; and
- Kwon and Rong-Ming (2005) applied FRFs and the Taguchi method.

In this chapter the stiffness matrices were updated by varying physical parameters such as the modulus of elasticity. The process of varying physical parameters was repeated until the Euclidean norm of the error that was caused by the integration of the experimental results and the finite-element model computed parameters was minimized. The physical parameter that minimizes the Euclidean norm of the error was used to obtain the new mass and stiffness matrices.

The equation of motion describing the dynamic characteristics of a structure may be obtained using Newton's second law of motion or by using a suitable energy principle to obtain an N degrees-of-freedom viscously damped system (Ewins, 1986):

$$[M]\{\ddot{x}\}+[C]\{\dot{x}\}+[K]\{x\}=\{F(t)\} \quad (8.1)$$

In Equation 8.1 $[M]$ is the mass matrix; $[C]$ is the damping matrix; $[K]$ is the stiffness matrix; $\{F(t)\}$ is the forcing function vector; $\{x\}$ is the displacement vector; $\{\dot{x}\}$ is the velocity vector; and $\{\ddot{x}\}$ is the acceleration vector.

The damping matrix was assumed to be proportional and was expressed in terms of mass and stiffness matrices as follows:

$$[C]=\alpha[M]+\beta[K] \quad (8.2)$$

Equation 8.2 may be rearranged by taking all terms to the left-hand side and may be rewritten as follows:

$$[M]\{\ddot{x}\}+[C]\{\dot{x}\}+[K]\{x\}-\{F(t)\}=\{0\} \quad (8.3)$$

If it is assumed that

$$\{x\}=\{X(\omega)\}e^{i\omega t} \quad (8.4)$$

and taking the appropriate derivatives, it then follows that:

$$\{\dot{x}\}=i\omega\{X(\omega)\}e^{i\omega t} \quad (8.5)$$

$$\{\ddot{x}\}=-\omega^2\{X(\omega)\}e^{i\omega t} \quad (8.6)$$

Substituting Equations 8.5 and 8.6 into Equation 8.3, the equation may be written in the frequency domain as follows:

$$\left(-\omega^2[M]+i\omega[C]+[K]\right)\{X(\omega)\}-\{F(\omega)\}=\{0\} \tag{8.7}$$

or

$$[B(\omega)]\{X(\omega)\}-\{F(\omega)\}=\{0\} \tag{8.8}$$

In Equation 8.8 $[B(\omega)]$ is the dynamic stiffness matrix of the structure. In Equation 8.7, $\{X(\omega)\}$ and $\{F(\omega)\}$ are measured quantities. The difficulty with Equation 8.7 is that FRFs are measured, instead of individual displacements and force. To solve this problem, the excitation was assumed to be white noise, and hence the vector $\{F(\omega)\}$ has a unit force magnitude at all frequencies, and the displacement is replaced by the FRFs.

If the measured FRFs are substituted into Equation 8.3, using $[M]$, $[C]$ and $[K]$ from the finite-element model, then there will be an error that will depend on the accuracy of the finite-element model. This system error vector may be introduced on the right hand side of Equation 8.8 and the resulting equation is:

$$[B(\omega)]\{X(\omega)\}-\{F(\omega)\}=\{\varepsilon(\omega)\} \tag{8.9}$$

Due to the cumbersome nature of investigating the elements of the error vector, the Euclidean norm (e) (Burden and Faires, 1993), which is the square root of the sum of the squares of the error vector elements, was used. If the error vector has zero elements, then e will be equal to zero. The equation for e is:

$$e=\left(\sum_{j=1}^{N}\left|\varepsilon(\omega_j)^2\right|\right)^{\frac{1}{2}} \tag{8.10}$$

where N is the number of frequency points.

The mass, stiffness, and damping matrices, amongst other things, depend on the area (A), density (ρ), Poisson ratio (v) and the modulus of elasticity (E) of each element. By varying one of these physical parameters, e can be minimized. The A, ρ, v, and E obtained by updating the finite-element model are thus known as *updating parameters.*

8.2.2 Modal Property Method (MPM)

The measured FRFs may be used to extract measured modal parameters, *i.e.*, the natural frequencies and mode shapes through using a process called modal analysis. *Modal analysis* is a process by which the dynamic properties of a mechanical structure under dynamic excitation can be studied using natural frequencies, mode shapes and damping. The advantage of using modal properties is that the method yields relatively greater immunity to noise in the data than does directly using the FRF data.

Some methods commonly used to achieve this conversion are the peak-picking method and the circle-fitting method (Friswell and Mottershead, 1995; Peeters and De Roeck, 2001). The *peak-picking method* uses the peak from the graph of the frequency-response function magnitude versus frequency to identify the modes. The *circle-fitting method* uses the geometrical properties of the Nyquist plot to identify the natural frequencies, mode shapes and damping information (Iglesias, 2000). Mode shapes have several special properties such as orthogonality. The MPM uses the orthogonality approach to update the finite-element model. The modified version of the eigenvalue equation will be developed in the next few sections.

Link and Weiland (2009) updated the finite-element model to correct chosen parameters of the model to better reflect experimental data. This was conducted by minimizing the distance between the analytical and experimental natural frequencies and mode shapes.

Bayraktar *et al.* (2009) successfully applied modal properties to update a finite-element model of a highway bridge. The experimental data were obtained through modal analysis of traffic loads from vibration data, collected from a girder and bridge deck. The peak-picking technique was used to identify modal data.

Gentile and Gallino (2008) used modal properties obtained from experimental data and merged with standard methods of system identification to update parameters of a finite-element model of a footbridge. The peak-picking method was used to extract modal data from ambient vibration data.

Camillacci and Gabriele (2005) applied modal data for updating of a finite-element model. The modal data were obtained by using a fractional polynomial method and an eigensensitivity model-updating method to adjust the structural parameters of a finite-element model. The method was successfully implemented to update a finite-element model of a three-floor framed structure.

Marwala and Sibisi (2005) applied modal properties to finite-element-model updating. The Bayesian method was introduced to solve the finite-element-updating problem. Standard deviations of the distributions were obtained through the use of the MCMC method to sample the distribution. The measured natural frequency data showed that for all the modes, the updated model was more accurate than the initial mode. It was found that the method significantly improved the accuracy of the finite-element models.

Jung and Park (2005) introduced the original idea of creating feedback loops to the conventional modal test setup and of using closed-loop natural frequency data for finite-element updating to overcome the problems with the usual technique based on a modal sensitivity matrix. A feedback loop was intended to alter the modal properties of the system and thereby dynamically control its characteristics.

El-Borgi *et al.* (2008) proposed a technique for the structural assessment of a reinforced-concrete bridge that was based on ambient vibration measurement and finite-element-model updating. An enhanced frequency domain decomposition method was implemented to extract the dynamic characteristics of the bridge and the finite-element model was updated to attain a sensible correspondence between experimental and numerical modal properties. The parameter chosen for the finite-element-model updating was the modulus of elasticity of the concrete in the elements of the finite-element model.

Yuen and Katafygiotis (2002) applied a Bayesian framework for the modal updating that allowed the user to gain the most probable updated parameters as well as their uncertainties. Other successful applications of modal data to finite-element updating include:

- Touat *et al.* (2007), who applied modal data and modified accelerated random searches for finite-element-model updating;
- Unger *et al.* (2006), who used modal data to update a finite-element model and used this for detecting damage in a prestressed-concrete beam;
- Gentile (2006) who introduced a finite-element-updating method that could be used to assess the safety of a bridge under the service load;
- Gabriele *et al.* (2004) for updating fern parameters; and
- Kharrazi *et al.* (2002), who used the modal data and finite-element-model updating procedure for detection of damage.

If the equation of motion for a dynamic system (Equation 8.1) is modified by setting $\{F(t)\}=\{0\}$, then the resulting free-vibration equation may be written as follows (Meirovitch, 2000):

$$[M]\{\ddot{x}\}+[C]\{\dot{x}\}+[K]\{x\}=\{0\} \tag{8.11}$$

Using differential equation theory, the solution may be assumed to be of the form (Meirovitch, 2000):

$$\{x\}=\{X(\omega)\}e^{i\omega t} \tag{8.12}$$

By differentiating appropriately, Equations 8.5 and 8.6 may be obtained. If these equations are substituted into Equation 8.11 then the resulting equation is:

$$(-\omega_j^2[M]+i\omega[C]+[K])\{X\}=\{0\} \tag{8.13}$$

By ignoring damping and thus setting $[C]=0$ and then setting $\{X\}=\{\psi\}$:

$$(-\omega_j^2[M]+[K])\{\psi\}=\{0\} \tag{8.14}$$

Equation 8.14 is a set of simultaneous equations with unknown ω_j and $\}=\{\psi\}_j$ for $j=1,\ldots,N$ where N is the number of degrees of freedom. From the theory of differential equations, to obtain the nontrivial equation, the determinant of the coefficient matrix must be equal to zero.

$$\left|-\omega_j^2[M]+i\omega[C]+[K]\right|=0 \tag{8.15}$$

If Equation 8.15 is expanded, a polynomial in terms of ω is obtained and is called the *characteristic equation* where

$$a_N\lambda^N + a_{N-1}\lambda^{N-1} + a_{N-2}\lambda^{N-2} + ... + a_1\lambda + a_0 = 0 \tag{8.16}$$

Here, $\lambda = \omega^2$.

The roots of the polynomial may be solved for ω_j and are called *eigenvalues* and correspond to the natural frequencies. Using ω_j the nonunique $\{\psi\}_j$ may be obtained. The solutions $\{\psi\}_j$ are called *eigenvectors* and correspond to mode shapes. Due to their nonuniqueness, eigenvectors $\{\psi\}_j$ have to be scaled. If one of the elements of the vector $\{\psi\}_j$ is given a specific value, then the vector has a unique value. The process of scaling the modal vector to obtain a unique vector is called *normalization.* When the mass matrix is premultiplied by a transpose of the normalized vector and postmultiplied by a normalized vector, then the modal mass is obtained. When the stiffness matrix is premultiplied by a transpose of the normalized vector and postmultiplied by a normalized vector, then the modal stiffness is obtained.

$$\{\psi\}_j^T[M]\{\psi\} = m_j \tag{8.17}$$

$$\{\psi\}_j^T[K]\{\psi\} = k_j \tag{8.18}$$

Here, T stands for transpose. If Equation 8.17 is scaled such that $m_j = 1$, then the vector $\{\psi\}_j$ becomes orthonormalized and may be replaced by the normalized mode shapes $\{\phi\}_j$ and the new equation is (Ewins, 1986):

$$\{\phi\}_j^T[M]\{\phi\} = 1 \tag{8.19}$$

$$\{\phi\}_j^T[K]\{\phi\} = \omega_j^2 \tag{8.20}$$

By setting $\{\psi\}_j = \{\phi\}_j$ in Equation 8.14 then

$$(-\omega_j^2[M] + [K])\{\phi\}_j = 0 \tag{8.21}$$

If Equation 8.21 is premultiplied by $\{\phi\}_j^T$ then the resulting expression is

$$\{\phi\}_j^T(-\omega_j^2[M] + [K])\{\phi\}_j = 0 \tag{8.22}$$

where T, again, stands for transpose. This equation may be rewritten as:

$$\omega_j^2\{\phi\}_j^T[M]\{\phi\}_j = \{\phi\}_j^T[K])\{\phi\}_j \tag{8.23}$$

When an experiment is conducted and modal properties of the system are extracted from the measured FRFs, it is often the case that the measured properties differ from the finite-element predictions (Friswell and Mottershead, 1995). The measured modal properties, the mass and the stiffness matrices from the finite-element model may be substituted in Equation 8.23. In this case, the left-hand side of Equation 8.23 is often different from the right-hand side. For this reason, Equation 8.23 may conveniently be written in terms of an error scalar as:

$$\varepsilon_j = \omega_j^2\{\phi\}_j^T[M]\{\phi\}_j - \{\phi\}_j^T[K])\{\phi\}_j \tag{8.24}$$

Equation 8.24 gives the error for a given measured mode j. Equation 8.24 can be written as follows to account for M modes:

$$e = \left\| \sum_{j=1}^{M} \left| \omega_j^2\{\phi\}_j^T[M]\{\phi\}_j - \{\phi\}_j^T[K])\{\phi\}_j \right| \right\| \tag{8.25}$$

Here, $|\;|$ stands for the absolute value and $\|\;\|$ stands for the Euclidean norm, which is shown in Equation 8.10.

Since the mass and stiffness matrices depend on the modulus of elasticity and other parameters, these parameters may be varied iteratively, thereby updating the mass and stiffness matrices until e is minimized. The physical parameters that give rise to the minimum e will give the updated finite-element model.

8.2.3 Multi-criteria Method (MCM)

The FRFM and MPM work well except for the fact that they usually give different results. To combine the information from both the modal properties and the FRFs, a method combining the frequency and the modal domains is used in this chapter. This method is essentially framed as a multi-criteria optimization problem. *Multi-criteria optimization* (Steuer, 1986; Sawaragi *et al.*, 1985; Das and Dennis, 1998; Deb, 2002) is the procedure for concurrently optimizing more than two objective functions that may be subjected to certain constraints.

If a multi-objective exercise is well formed then there should not be one answer that concurrently minimizes any of objective function to its fullest. Schlune *et al.* (2009) introduced multi-response objective functions that permitted hybridizing different types of measurements to update a finite-element model of a bridge. Wang *et al.* (2009) applied a multi-objective concurrent method for a civil infrastructure to concurrently analyze damage analysis and assess the state.

Christodoulou *et al.* (2008) applied multi-criteria optimization for model updating and variability estimation using Pareto optimal models. Their multi-criteria identification technique model updating was based on modal residuals. The technique resulted in multiple Pareto optimal models that are consistent with the experimentally measured modal data. When the model updating techniques were compared it was observed that the Pareto optimal models and the related response and reliability estimations may differ significantly because of the model class and the measurement errors.

D'Ambrogio and Fregolent (1998) introduced procedures for model updating based on inverse sensitivity and another one based on multi-objective optimization. The results obtained demonstrated that both methods perform well with drive point FRFs. Perera and Ruiz (2008) formulated a multi-objective function that incorporated modal flexibilities and a damage-location criteria and used this to detect damage on a bridge. Other applications of the multi-criteria method include:

- De Castro *et al.* (2008) who applied a multi-objective genetic algorithm to identify the unbalanced mode for rotating machinery;
- Papadimitriou and Ntotsios (2008) for system integration; and
- Pascual *et al.* (1998) for updating industrial models.

The errors obtained in Equations 8.10 and 8.24 may be combined using equal weighting functions to obtain the equation:

$$
\begin{aligned}
e = & \left\| \sum_{j=1}^{M} \left| \omega_j^2 \{\phi\}_j^T [M]\{\phi\}_j - \{\phi\}_j^T [K])\{\phi\}_j \right| \right\| \ldots \\
& + \left\| \sum_{j=1}^{N} \left\| [B(\omega_j)]\{X(\omega_j) - \{F(\omega_j)\} \right\| \right\|
\end{aligned}
\tag{8.26}
$$

It was discovered that the method works better if a normalization factor is introduced into Equation 8.25:

$$
\begin{aligned}
e = & \frac{1}{f_1} \left\| \sum_{j=1}^{M} \left| \omega_j^2 \{\phi\}_j^T [M]\{\phi\}_j - \{\phi\}_j^T [K])\{\phi\}_j \right| \right\| \ldots \\
& + \frac{1}{f_2} \left\| \sum_{j=1}^{N} \left\| [B(\omega_j)]\{X(\omega_j) - \{F(\omega_j)\} \right\| \right\|
\end{aligned}
\tag{8.27}
$$

here,

$$f_1 = \left\| \sum_{j=1}^{M} \left| \omega_j^2 \{\phi\}_j^T [M]\{\phi\}_j - \{\phi\}_j^T [K])\{\phi\}_j \right| \right\|_0 \tag{8.28}$$

and

$$f_2 = \left\| \sum_{j=1}^{N} \left\| [B(\omega_j)]\{X(\omega_j) - \{F(\omega_j)\} \right\| \right\|_0 \tag{8.29}$$

In Equations 8.28 and 8.29 the subscript 0 indicates the parameters of the initial estimation of the finite-element parameters. The parameters used for finite-element updating in this chapter were the moduli of elasticity of each element.

8.3 Optimization

To minimize Equations 8.10, 8.24 and 8.26 a hybrid optimization method that combined particle-swarm optimization and the Nelder–Mead simplex method was implemented as outlined in Chapter 7. As indicated before, the particle-swarm-optimization method was used because of its global search capabilities, while the Nelder–Mead search algorithm described in detail in Chapter 2 was used because of its local search capabilities. These optimization methods have been used before to solve complex problems (Marwala, 2005; Mthembu *et al.*, 2010, Marwala, 2009). The algorithm implemented is outlined below:

1. Initialize a population of particles' positions and velocities. The positions of the particles must be randomly distributed in the updating parameter space.
2. Calculate the velocity for each particle in the swarm.
3. Update the position of each particle.
4. Repeat Steps 2 and 3 until convergence is achieved.
5. Pick the three best results from the swarm that were identified in Step 3 and use this to form a simplex vertex.
6. For each step in the optimization procedure, replace the vertex with the worst fitness measure as defined via the objective function by the new vertex.
7. Establish the coordinates of the new vertex by reflecting the old vertex's point about the outstanding vertices.
8. If the fitness measure of the current vertices is lower than the preceding one, the dimensions of the simplex are minimized. If not, they are enlarged.

This process in Steps 6 to 8 is continued until the functional evaluation values of the vertices become similar.

8.4 Example 1: Simple Beam

The aluminum beam shown in Chapter 2 was used to apply the MPM, FRFM and MCM methods described earlier, for finite-element updating. More information on this can be found in Marwala (1997). The beam was tested freely suspended using elastic rubber bands. The beam was excited using an electromagnetic shaker and the response was measured using an accelerometer. The beam was also modeled using a finite-element model, as explained in the previous chapters, and it was divided into 11 elements. It was excited at various positions and the acceleration was measured at 10 different positions.

A set of 10 frequency-response functions were calculated and a roving accelerometer was used for the testing. In this chapter the moduli of elasticity of these elements were used as updating parameters.

On implementing the hybrid particle-swarm optimization for finite-element-model updating the following attributes were used: population size of 50, c_1 of 0.05, c_2 of 0.01 and $w = 0.002$.

The particle-swarm optimization and genetic algorithm were run for 200 generations. An arithmetic crossover probability of 40% and nonuniform mutation probability of 0.5% were implemented.

Table 8.1 Results showing measured frequencies of the beam, the initial frequencies and the frequencies obtained when the finite-element model of a beam was updated using the MPM, FRFM and MCM methods

Modes	Measured frequency (Hz)	Initial frequency (Hz)	The MPM (Hz)	The FRFM (Hz)	The MCM (Hz)
1	41.5	42.3	42.1	41.9	41.6
2	114.5	117.0	116.8	111.2	115.4
3	224.5	227.3	222.3	226.7	225.7
4	371.6	376.9	368.8	374.9	371.1

When a modulus of elasticity of 7×10^{10} N m^{-2} was assumed, the error between the first measured natural frequency and that from the initial finite-element model was 1.9%. When the MPM was used for finite-element-model updating this error was reduced to 1.4%, whereas using the FRFM reduced this error to 1.0% and using the MCM reduced it to 1.7%.

The error between the second measured natural frequency and that of the initial model was 2.2%. When the MPM was used, this error was reduced to 2.0%; when the FRFM was used, it was increased to 2.9%, while using the MCM it was reduced to 0.7%.

The error in the third natural frequencies between the measured data and the initial finite-element model was 1.2%. When the MPM was used, this error was reduced to 1.0%; using the FRFM reduced the error to 1.0% and using the MCM reduced it to 1.1%.

The error between the fourth measured natural frequency and that of the initial model was 1.4%. When the MPM was used, this error was reduced to 0.8%; using the FRFM reduced this error to 1.3% and using the MCM it was reduced to 0.4%.

Overall, the MCM gives the best results, followed by the MPM. On average, all three methods improved the finite-element models when compared to the average error between the initial finite-element model and the measured data.

The updated models implemented were also validated on the mode shapes they predicted. To make this assessment possible, the modal assurance criteria (MAC) was used (Allemang and Brown, 1982), and the results are shown in Table 8.2.

The mean of the diagonal of the MAC vector was used to compare the mode shapes predicted by the updated and initial finite-element models with the measured mode shapes.

An average value of 1.0 indicates that the mode shapes are properly correlated. The average MAC calculated between the mode shapes of the initial finite-element model and the measured mode shapes was 0.9986. When the average MAC was calculated between the measured data and data obtained from the updated finite-element models, it was observed that the MPM gave marginally better results, followed by the MCM and then the FRFM.

Table 8.2 Beam results showing the MAC calculated between measured mode shapes and the initial finite-element model, the MPM, FRFM and MCM

Method	MAC
Initial model	0.9986
MPM	0.9989
FRFM	0.9987
MCM	0.9988

8.5 Example 2: Unsymmetrical H-shaped Structure

The unsymmetrical H-shaped aluminum structure shown in Chapter 2 was also used to validate the MPM, FRFM and MCM techniques. This structure was previously used by Marwala (1997). The structure was excited using an electromagnetic shaker and the response was measured using an accelerometer. The structure was divided into 12 elements. It was excited and the acceleration was measured at 15 positions. The structure was tested freely suspended, and a set of 15 frequency-response functions were calculated. A roving accelerometer was used for the testing. The mass of the accelerometer was found to be negligible compared to the mass of the structure.

As in the previous example, the finite-element model was constructed using the Euler–Bernoulli beam elements. The finite-element model contained 12 elements. As in the previous example, the moduli of elasticity of these elements were used as updating parameters, restricted to fall in the interval 6×10^{10} to 8×10^{10} N m^{-2}. On implementing the hybrid particle-swarm optimization for finite-element-model updating the following parameters were selected:

- population = 50;

- $c_1 = 0.05$;
- $c_2 = 0.01$; and
- $w = 0.002$.

The particle-swarm optimization and genetic algorithm were run for 200 generations. Arithmetic crossover probability was 40% and a nonuniform mutation probability of 0.5% was implemented.

The results obtained when the MPM, FRFM and MCM techniques were used for finite-element-model updating are shown in Table 8.3. Table 8.3 shows the measured and initial natural frequencies as well as the natural frequencies obtained by the MPM-, FRFM- and MCM-updated finite-element models. When a modulus of elasticity of 7×10^{10} N m^{-2} was assumed, the error between the first measured natural frequency and that from the initial finite-element model was 4.3%. When the MPM method was used for finite-element-model updating, this error was reduced to 3.2%, while the FRFM increased this error to 4.6% and the MCM approach reduced the error to 3.5%.

Table 8.3 Results from an unsymmetrical H-shaped structure showing measured frequencies, the initial frequencies and the frequencies obtained when the finite-element model was updated using the MPM, FRFM and MCM techniques

Modes	Measured frequency (Hz)	Initial frequency (Hz)	The MPM (Hz)	The FRFM (Hz)	The MCM (Hz)
1	53.9	56.2	52.2	56.3	52.0
2	117.3	127.1	119.3	116.8	119.1
3	208.4	228.4	212.1	213.9	211.3
4	254.0	261.4	257.7	250.3	256.6
5	445.1	452.4	443.9	448.1	443.9

The error between the second measured natural frequency and that from the initial model was 8.4%. When the MPM method was used, the error was reduced to 1.7%, while using the FRFM technique gave an error of 0.4% and the MCM reduced this error to 1.5%.

The error in the third natural frequencies between the measured data and the initial finite-element model was 9.6%. When the MPM method was used, this error was reduced to 1.8%, while the FRFM technique reduced this error to 2.6% and using the MCM it was reduced to 2.9%.

The error between the fourth measured natural frequency and that of the initial model was 3.7%. When the MPM method was used, this error was reduced to 1.4%, while the FRFM reduced this error to 1.4 and using the MCM it was reduced to 1.0%.

The error between the fifth measured natural frequency and that from the initial model was 1.6%. When the MPM method was used, this error was reduced to 0.3%, while the FRFM reduced it to 0.7% and the MCM reduced it to 0.3%.

Overall, the genetic programming MCM gave the best results, followed by the MPM method. As in the previous example, the updated models were validated on the mode shapes they predicted using the MAC. The results appear in Table 8.4. Table 8.4 shows that the MPM gave the best results, followed by the MCM.

Table 8.4 Results of the unsymmetrical H-shaped structure showing measured frequencies, the initial frequencies and the frequencies obtained when the finite-element model was updated using the MPM. FRFM and MCM methods

Method	MAC
Initial model	0.8394
MPM	0.8513
FRFM	0.8497
MCM	0.8501

8.6 Conclusion

In this chapter a multiple criteria method (MCM) was tested on a simple beam and an irregular H-shaped structure. The MCM is a multi-criteria method that minimizes the Euclidean norm of the error vector obtained when the modal properties and the frequency-response functions (FRFs) are used. The MCM was compared to methods when the modal properties were used in isolation (MPM) and when the FRFs were used in isolation. The results obtained demonstrated that the MCM gave the best performance in updating the finite-element models, followed by the MPM.

8.7 Future Work

This chapter introduced the MCM for finite-element-model updating and compared it with using the MPM and the FRFM. These procedures were implemented by using hybrid optimization methods that combined the Nelde–Mead simplex method and particle-swarm optimization. For further work, other optimization methods should be used for finite-element-model updating. The MCM was formulated by combining the modal-domain data with the frequency-domain data. For future work, time–frequency domain data such as the wavelet transform should be incorporated into the MCM.

References

Allemang RJ, Brown DL (1982) A Correlation Coefficient for Modal Vector Analysis. In: Proc of the 1st Int Modal Analysis Conf:01–18

Arora V, Singh SP, Kundra TK (2009) Finite Element Model Updating with Damping Identification. J of Sound and Vib 324:1111–1123

Asma F, Bouazzouni A (2005) Finite Element Model Updating Using FRF Measurements. Shock and Vib 12:377–388

Bayraktar A, Altunişik AC, Sevim B, Türker T (2009) Finite Element Model Updating of Kömürhan Highway Bridge. Tech J of Turkish Chamber of Civil Eng 20:4675-700

Burden RL, Faires JD (1993) Numerical Analysis.PWS Publishing, Massachusetts

Camillacci R, Gabriele S (2005) Mechanical Identification and Model Validation for Shear-type Frames. Mech Syst and Signal Process 19:597–614

Christodoulou K, Ntotsios E, Papadimitriou C, Panetsos P (2008) Structural Model Updating and Prediction Variability Using Pareto Optimal Models. Comput Methods in Appl Mech and Eng 198:138–149

D'Ambrogio W, Fregolent A (1998) New Figures of Merit for Non-modal Test-analysis Correlation. In: Proc of the 23rd Intl Conf on Noise and Vib Eng, ISMA:393–400

D'Ambrogio W, Zobel PB (1990) Damage Detection in Truss Structures Using a Direct Updating Technique. ISMA19 Tools for Noise and Vib Analysis II

Das A, Dennis JE (1998) Normal-Boundary Intersection: A New Method for Generating the Pareto Surface in Nonlinear Multicriteria Optimization Problems. SIAM J on Optim 8:631–657

Dascotte E, Strobbe J (1999) Updating Finite Element Model Using FRF Correlation Functions. In: Proc of the Intl Modal Analysis Conf

De Castro HF, Cavalca KL, De Camargo LWF (2008) Multi-objective Genetic Algorithm Application in Unbalance Identification for Rotating Machinery. In: Proc of the 9th Intl Conf on Vib in Rotating Machin 2:885–897

Deb K (2002) Multi-Objective Optimization Using Evolutionary Algorithms. Wiley, New York

El-Borgi S, Neifar M, Cherif F, Choura S, Smaoui H (2008) Modal Identification, Model Updating and Nonlinear Analysis of a Reinforced Concrete Bridge. J of Vib and Contr 14:511–530

Esfandiari A, Bakhtiari-Nejad F, Rahai A, Sanayei M (2009) Structural Model Updating Using Frequency Response Function and Quasi-linear Sensitivity Equation. J of Sound and Vib 326:557–573

Ewins DJ (1986) Modal Testing: Theory and Practice. Research Studies Press, Letchworth

Friswell MI, Mottershead JE (1995) Finite Element Model Updating in Structural Dynamics. Kluwer Academic Publishers Group, Norwell

Gabriele S, Valente C, Brancaleoni F (2004) Interval Analysis for Updating Fern Parameters Using Uncertain Experimental Data. In: Proc of the Intl Conf on Noise and Vib Eng, ISMA:3065–3077

Gentile C (2006) Modal and Structural Identification of a R.C. Arch Bridge. Struct Eng and Mech 22:53–70

Gentile C, Gallino N (2008) Ambient Vibration Testing and Structural Evaluation of an Historic Suspension Footbridge. Adv in Eng Softw 39:356–366

Hemez FM, Doebling SW (1999) Test-analysis Correlation and Finite Element Model Updating for Nonlinear, Transient Dynamics. In: Proc of the Intl Modal Analysis Conf – IMAC:1501–1510

Iglesias AM (2000) Investigating Various Modal Analysis Extraction Techniques to Estimate Damping Ratio. Master of Sci Thesis, Virginia State Univ and Polytech Inst

Jung H, Park Y (2005) Model Updating Using the Closed-loop Natural Frequency. J of Guid, Contr, and Dyn 28:20–29

Kerschen G, Golinval JC (2004) A Two-step Methodology for the Generation of Accurate Finite Element Models of Nonlinear Mechanical Systems. Am Soc of Mech Eng, Des Eng Div (Publ) DE 117:313–320

Kharrazi MHK, Ventura CE, Brincker R, Dascotte E (2002) A Study on Damage Detection Using Output-only Modal Data. In: Proc of SPIE - The Intl Soc for Optical Eng 4753 II:1199–1205

Kwon KS, Rong-Ming L (2005) Robust Finite Element Model Updating Using Taguchi Method. J of Sound and Vib 280:77–99

Lin RM, Zhu J (2007) Finite Element Model Updating Using Vibration Test Data Under Base Excitation. J of Sound and Vib 303:596–613

Link M, Weiland M (2008) Damage Identification by Computational Model Updating. In: Proc of the 9th Biennial Conf on Eng Syst Des and Anal 2:683–693

Link M, Weiland M (2009) Damage Identification by Multi-model Updating in the Modal and in the Time Domain. Mech Syst and Signal Process 23:1734–1746

Lu Y, Tu Z (2004) A Two-level Neural Network Approach for Dynamic FE Model Updating Including Damping. J of Sound and Vib 275:931–952

Marwala T (1997) A Multiple Criteria Updating Method for Damage Detection on Structures. Master's Thesis, University of Pretoria

Marwala T (2002) Finite Element Updating Using Wavelet Data and Genetic Algorithm. J of Aircr 39:709–711

Marwala T (2005) Finite Element Model Updating Using Particle Swarm Optimization. Intl J of Eng Simul 6:25–30

Marwala T (2009) Computational Intelligence for Missing Data Imputation, Estimation and Management: Knowledge Optimization Techniques. IGI Global, New York

Marwala T, Heyns PS (1998) Multiple Criteria Method for Determining Structural Damage. AIAA J 36:1494–501

Marwala T, Sibisi S (2005) Finite Element Model Updating Using Bayesian Framework and Modal Properties. J of Aircr 42:275–278

Meirovitch L (2000) Fundamentals of Vibrations. McGraw-Hill Companies, Singapore

Mthembu L, Marwala T, Friswell MI, Adhikari S (Accepted) Finite Element Model Selection Using Particle Swarm Optimization. 2010 Intl Modal Analysis Conf

Nobari AS, Farjoo MA (2004) Comparison Between Dedicated Model Updating Methods and Hybrid Method. J of Aircr 41:999–1004

Papadimitriou C, Ntotsios E (2008) Optimization Algorithms for System Integration. In: Proc of the 3rd Intl Conf on Smart Mater, Struct and Syst:514–523

Pascual R, Golinval JC, Berthillier M, Despres T (1998) Updating Industrial Models Under a General Optimization Environment. In: Proc of the Intl Modal Analysis Conf – IMAC, 2:1326–1332

Peeters B, De Roeck G (2001) Stochastic System Identification for Operational Modal Analysis: A Review. J of Dyn Syst, Meas and Control 123:659–668

Perera R, Ruiz A (2008) Damage Detection in Large Scale Structures Using a Two Stage Procedure. In: Proc of the 4th European Workshop on Struct Health Monit:1279–1286

Ren Y, Beards CF (1995) Identification of Joint Properties of a Structure Using FRF Data. J of Sound and Vib 186:567–587

Sadr MH, Astaraki S, Salehi S (2007) Improving the Neural Network Method for Finite Element Model Updating Using Homogenous Distribution of Design Points. Arch of Appl Mech 77:795–807

Sawaragi Y, Nakayama H, Tanino T (1985) Theory of Multiobjective Optimization. Academic Press Inc, Orlando

Schlune H, Plos M, Gylltoft K (2009) Improved Bridge Evaluation Through Finite Element Model Updating Using Static and Dynamic Measurements. Eng Struct 31:1477–1485

Steuer RE (1986) Multiple Criteria Optimization: Theory, Computations, and Applications. John Wiley & Sons, New York

Tong ZP, Zhang Y, Shen RY, Hua HX (2005) Submarine Finite Element Model Updating Method Based on Frequency Response Functions of Vibration. J of Shanghai Jiaotong Univ 39:1847–1850

Touat N, Pyrz M, Rechak S (2007) Accelerated Random Search Method for Dynamic FE Model Updating. Eng Comp 24:450–472

Unger JF, Teughels A, De Roeck G (2006) System Identification and Damage Detection of a Prestressed Concrete Beam. J of Struct Eng 132:1691–1698

Wang C, Li Z, Yin A (2009) Multi-objective Concurrent Approaching of Simulating for Civil Infrastructure. J of Southeast Univ (Natl Sci Ed) 39:78–84

Yuen KV, Katafygiotis LS (2002) Bayesian Modal Updating Using Complete Input and Incomplete Response Noisy Measurements. J of Eng Mech 128:340–350

Ziaei-Rad S (2005) Finite Element Model Updating of Rotating Structures Using Different Optimisation Techniques. Iranian J of Sci and Technol, Trans B: Eng 29:569–585

Chapter 9

Finite-element-model Updating Using Artificial Neural Networks

Abstract. This chapter implements Bayesian neural networks for finite-element-model updating. This method was tested on a simple beam and an unsymmetrical H-shaped structure and compared with an implementation that was based on the response-surface method. It was observed, on average, that the Bayesian neural-network approach gave more accurate results than the response-surface method did.

Keywords: neural networks, Bayesian neural networks, stochastic dynamics model, hybrid Monte Carlo

9.1 Introduction

As indicated before, finite-element-model updating is a process of tuning a finite-element model so that it can better predict the observed data from the physical structure being modeled (Friswell and Mottershead, 1995). The finite-element model of a physical object approximates this object. Finite-element-model updating is the process of identifying a better approximation of the physical object than the initial finite-element model. This procedure is essentially an optimization problem, where the design variables are the parameters of the finite-element model that are deemed to be in doubt and, therefore, need to be updated. The objective function characterizes the distance between the finite-element model predictions and those from measurements.

Many methods have been used for finite-element-model updating; some of these involve machine-learning techniques. *Machine learning* is a subject that deals with the design, development and implementation of techniques that permit computers to learn based on data (Alpaydın, 2004). The main objective of machine-learning is to autonomously learn the identification of complicated patterns and take intelligent decisions based on data. For this reason, machine-learning is intimately linked to disciplines such as probability theory, statistics and computational intelligence. Some successful applications of machine-learning include:

- fault identification in mechanical systems (Marwala and Hunt, 1999);
- natural language processing (Nelwamondo *et al.*, 2006);
- recommender systems (Marivate *et al.*, 2008);
- medical diagnosis (Mohamed *et al.*, 2006);
- bioinformatics (Mohamed *et al.*, 2007);
- stock-market analysis (Patel and Marwala, 2006);
- classifying gene expression (Perez *et al.*, 2008);
- object recognition (Machowski and Marwala, 2004); and
- game playing (Hurwitz and Marwala, 2005).

Machine-learning methods are classified into categories based on the objective of the technique (Bhagat, 2005). These methods include:

- *Supervised learning*: this identifies a mathematical function that maps inputs to outputs (Vilakazi and Marwala, 2006). Examples of these methods include multi-layer perceptron and support vector machines (Msiza *et al.*, 2008).
- *Unsupervised learning*: this models a set of inputs where the outputs are not available (Bishop, 1995).
- *Reinforcement learning*: this learns how to take action for a particular observation in the environment by ensuring that each action has an impact in the environment and the environment offers feedback in the form of rewards and punishment to direct the learning technique (Hurwitz and Marwala, 2005).

This chapter implements supervised learning methods in the form of Bayesian neural networks (Marwala, 2004) for finite-element-model updating. A Bayesian approach to finite-element-model updating has been successfully implemented in the past (Marwala and Sibisi, 2005; Mthembu *et al.*, 2009). Bayesian neural networks were used for finite-element-model updating because they offer the following attributes:

- resistance to noise in the data;
- the ability to avoid complex optimization processes; and
- the ability to operate with an incomplete number of experimentally measured degrees of freedom and modes.

Zapico *et al.* (2008) applied neural networks for the finite-element-model updating of a small steel frame. The network was constructed to map the natural frequencies to the updating parameters. The updating procedure was improved through using a regressive method. The results demonstrated that the updated finite-element model correctly replicates the low modes that were experimentally identified. He *et al.* (2008) applied a hybrid of genetic algorithms and neural networks for the model-updating method. The model-updating technique, based on a neural network, was improved using a uniform design method to create samples to decrease the number of samples in the procedure for structural-damage simulation. A genetic algorithm was used to optimize the first weight of the back-propagation in the network. The structure was broken into multilayer substructures, and the updating process was implemented step by step.

Fei *et al.* (2005) used neural networks for the finite-element-model updating of nonlinear structures. This procedure was applied to the process of updating a beam with nonlinear elements. Frequency responses were used to train the neural network and the results demonstrated the success of the proposed method.

Fei and Zhang (2004) applied radial basis functions for finite-element-model updating. Design parameters and features were treated as independent variables to be mapped using a radial basis functional neural network. Their updating method was applied to update an aircraft model. The results obtained showed errors of the design parameters of the current technique to be less than 2%, while the errors in the modal frequencies were less than 1%.

Lu and Tu (2004) applied a two-stage neural-network approach for finite-element-model updating, in which both the structural parameters and the damping ratios were updated. In the first stage, the model was updated under the assumption that the model was free of damping and the structural parameters were updated through using the natural and anti-resonance frequencies as the response data. The second-stage updating process concerned the damping ratios and used the integrals of the frequency-response function. It was observed that the proposed technique was resistant to noise.

Chang *et al.* (2002) used neural networks to choose samples for model updating. They examined the use of orthogonal arrays for the sample selection. Their results showed that the orthogonal arrays technique could considerably decrease the number of training samples with no influence on the accuracy of the neural-network prediction.

Levin *et al.* (2000) improved a neural network that was used for model updating. This improvement was aimed at ensuring that the neural network was resistant to noise. This was achieved by training the neural network with noise. Chang *et al.* (2000) applied adaptive neural networks for the model updating of structures. The neural network was first trained offline using training data acquired from finite-element analyses and including modal parameters as inputs and structural parameters as outputs. This neural network was afterward adaptively retrained online during the model-updating procedure to reduce the error between the measured and the predicted modal parameters. This procedure was applied to update a suspension-bridge model and tested numerically and experimentally. The results point out that it is possible to decrease the errors between the measured and the predicted frequencies from a maximum of 17 to 7% for the first eight vertical modes.

Levin and Lieven (1998) applied neural networks for finite-element-model updating. This updating technique was implemented on a simple simulated model, with and without the presence of noise, with promising results. An additional benefit of this updating technique was observed to be the capacity to operate with a limited number of experimentally measured degrees of freedom and modes.

Zhu and Zhang (2009) applied support vector machines for finite-element-model updating. In this study, features were treated as independent variables, while design parameters were treated as dependent variables. This method was applied to update a finite-element model of an aircraft, and the results demonstrated that the errors of design parameters and modal frequencies, respectively, were less than 2 and 1%.

Tan *et al.* (2009) applied wavelet analysis and support vector machines for the finite-element-model updating of structures. The element's stiffness matrix was represented by a semirigid node with a fixity coefficient. The computed acceleration response signal was decomposed by wavelet analysis. A support vector machine was set up by training samples that were acquired by extracting the wavelet signal component-node energy. The wavelet input data were mapped to the fixity coefficient using support vector machines. The simulation results gave a good updated finite-element model.

Liu *et al.* (2009) applied fuzzy theory to finite-element-model updating. The relationships between the modal parameters and design variables were modeled as fuzzy variables and using this, a fuzzy model-updating technique was presented. The technique was implemented on a real concrete bridge for which a physically important model was identified.

Further examples include Ait-Salem Duque *et al.* (2007) who applied a fuzzy finite-element method for choosing the initial conditions in the model updating of welded joints, as well as Atalla (1999) who applied neural networks for model updating.

9.2 Bayesian Neural Networks

In this chapter, neural networks are used for finite-element-model updating. In particular, this updating procedure uses Bayesian neural networks that are trained using a hybrid Monte Carlo technique. Maiti and Tiwari (2009) applied a hybrid Monte-Carlo-trained neural network for rock boundaries identification, while Gnewuch (2009) studied the probabilistic results for the inconsistency of the hybrid-Monte Carlo sequence. Desgranges and Delhommelle (2009) used a hybrid Monte Carlo to study the phase equilibria of molecular fluids and applied this technique to benzene and n-alkanes.

Cheung and Beck (2009) applied a hybrid Monte Carlo for Bayesian model updating and applied this technique to structural dynamic models with many uncertain parameters. The efficiency of the method for Bayesian model updating of structural dynamic models with many uncertain parameters was successfully demonstrated with simulated data for a ten-storey building that had 31 model parameters to be updated. Bogaerts (2009) applied a hybrid Monte Carlo method in a fluid model for studying the impact of nitrogen accumulation on argon glow discharges, while Kułak (2009) applied a hybrid Monte Carlo method to the donor–mediator–acceptor system in uniaxially stretched polymer film. Other successful applications of the hybrid Monte Carlo method include:

- a molecule-based magnet (Henelius and Fishman, 2008);
- financial time-series analysis (Takaishi, 2008);
- molecular simulation (Weber *et al.*, 2008); and
- crystallization (Desgranges and Delhommelle, 2008).

A *neural network* is a computational tool that is used to model the relationship between some input and output parameters (Leke *et al.*, 2006). In this chapter, neural networks are used to relate updating parameters to natural frequencies and are viewed as parameterized regression models that make probabilistic assumptions about the data. The probabilistic outlook of these models is facilitated by the use of a Bayesian framework. Learning algorithms were viewed as methods for finding parameter values that look probable in the light of the data. The learning process is usually conducted by dividing the data into training, validation and testing sets. This is done to select the model and to ensure that the trained network is not biased towards the training data it has seen. Another way of achieving this is by the use of a regularization framework, which comes naturally from the Bayesian formulation discussed in detail in this chapter. There are several types of neural-network procedures; these include:

- the multi-layer perceptron; and
- the radial basis functions (Bishop, 1995).

The multi-layer perceptron is used in this chapter because it provides a distributed representation with respect to the input space due to the crosscoupling between hidden units. In this chapter, the multi-layer perceptron architecture contains a tangent basis function in the hidden units and linear functions in the output units (Bishop, 1995). This network architecture contains hidden units and output units and has one hidden layer. The relationship between the output (*y*), representing the natural frequencies, and input (i), representing the updating parameters may be written as follows (Bishop, 1995):

$$y_k = f_{outer}\left(\sum_{j=1}^{M} w_{kj}^{(2)} f_{inner}\left(\sum_{i=1}^{d} w_{ji}^{(1)} x_i + w_{j0}^{(1)}\right) + w_{k0}^{(2)}\right) \tag{9.1}$$

Here, $w_{ji}^{(1)}$ and $w_{ji}^{(2)}$ indicate weights in the first and second layer, respectively, going from input i to hidden unit j; M is the number of hidden units; D is the number of output units, and $w_{j0}^{(1)}$ indicates the bias for the hidden unit j; and f_{outer} and f_{inner} arc, rcspcctivcly, a logistic outcr activation function and a hypcrbolic tangent activation function.

The process of training a neural network identifies the weights in Equation 9.1. An objective function must be chosen to identify these weights. If the training set $D = \{\{x_k\},\{y_k\}\}_{k=1}^{N}$ is used and assuming that the target vector $\{y\}$ is sampled independently given the input vector $\{x_k\}$ and the weight parameters vector, $\{w_{kj}\}$ the objective function, E, may thus be written by using the sum of squares of errors as (Bishop, 1995):

$$E = \beta \sum_{n=1}^{N} \sum_{k=1}^{K} \{y_{nk} - t_{nk}\}^2 + \frac{\alpha}{2} \sum_{j=1}^{W} w_j^2 \tag{9.2}$$

The sum of squares of error function was chosen because it has been established to be suited to regression problems (Bishop, 1995). In Equation 9.2, n is the index for the training pattern, β is the data contribution to the error and k is the index for the output units. The second term is the regularization parameter and it penalizes weights of large magnitudes (Bishop, 1995). This regularization parameter is called the *weight decay* and its coefficient α determines the relative contribution of the regularization term on the training error. This regularization parameter ensures that the mapping function is smooth and therefore the training process does not over-train and thus become biased towards the training data set. Including a regularization parameter has been found to give significant improvements in network generalization. The regularization component, in Equation 9.2 may be viewed as prior information within the Bayesian framework and will be elaborated on further in this chapter.

If α is too high, then the regularization parameter over-smoothes the network weights, thereby giving inaccurate results. If α is too small then the effect of the regularization parameter is negligible and, unless other measures are implemented to control the complexity of the model such as the early stopping method or cross-validation method (Bishop, 1995), then the trained network becomes too complex and thus performs poorly on the validation set.

Before network training is performed, network architecture needs to be constructed by choosing the number of hidden units, M. If M is too small, the neural network will be insufficiently flexible and will give a poor generalization of the data because of high bias. However, if M is too large, then the neural network will be unnecessarily flexible and will give poor generalization due to a phenomenon known as over-fitting caused by high variance.

In this chapter, the number of hidden nodes is chosen through trial and error. The problem of identifying the weights vector $\{w\}$ may be posed in Bayesian form as follows (Bishop, 1995; Marwala and Sibisi, 2005):

$$P(\{w\} \mid D) = \frac{P(D \mid \{w\})P(\{w\})}{P(D)} \tag{9.3}$$

In Equation 9.3 the parameter $P(\{w\})$ is the probability distribution function of the weight-space in the absence of any data, also known as the prior distribution, and $D = (x_1,\ldots,x_N,\ y_1,\ldots,y_N)$ is a matrix containing the input and output data. The quantity $P(\{w\}|D)$ is the posterior probability distribution after the data have been seen and $P(D|\{w\})$ is the likelihood function. Equation 9.3 may be expanded using Equation 9.2 to give (Marwala, 2004):

$$P(\{w\} \mid [D]) = \frac{1}{Z_s}\exp\left(-\beta\sum_{n}^{N}\sum_{k}^{K}\{y_{nk} - t_{nk}\}^2 - \frac{\alpha}{2}\sum_{j}^{W} w_j^2\right) \tag{9.4}$$

In Equation 9.4,

$$Z_S(\alpha,\beta) = \int \exp(-\beta E_D - \alpha E_W)\, dw$$
$$= \left(\frac{2\pi}{\beta}\right)^{N/2} + \left(\frac{2\pi}{\alpha}\right)^{W/2} \tag{9.5}$$

In Equation 9.4, the optimal weight vector corresponds to the maximum of the posterior distribution that was identified using the scaled conjugate gradient method. The distribution in Equation 9.4 is a canonical distribution (Haykin, 1999). Training the network using a Bayesian approach automatically penalizes highly complex models and therefore can select an optimal model without applying independent methods such as crossvalidation (Bishop, 1995), and also gives a probability distribution of the output of the networks that can be used to assess the reliability of the estimated updated parameters.

There are many methods that have been applied to solve Equation 9.4 including, most recently, a method proposed by Marwala (2007) that samples the weight space using genetic programming and others (Vivarelli and Williams, 2001). Further details on Bayesian neural-network training and the application of this in a mechanical system may be found in Neal (1992–1994) and Marwala (2001).

In this chapter, the method of sampling through a posterior distribution of weights, described in Equation 9.4, called the *hybrid Monte Carlo method* (Neal, 1994) is reviewed and then applied. Distributions of this nature (Equation 9.4) have been studied extensively in statistical mechanics. In statistical mechanics, macroscopic thermodynamic properties are derived from the state space, *e.g.*, position and momentum of microscopic objects such as molecules. The number of degrees of freedom that these microscopic objects have is enormous, so the only way to solve this problem is to formulate it in a probabilistic framework.

The hybrid Monte Carlo method, that uses the gradient of the error that is calculated using back-propagation, is used in this chapter to identify the posterior probability of the weight vectors, given the training data. The use of the gradient ensures that the simulation samples through the regions of higher probabilities and thus increases the time it takes to converge on a stationary probability distribution function. This technique is viewed as a form of a Markov chain with transition between states achieved by alternating the "stochastic" and "dynamic" moves. The "stochastic" moves allow the algorithm to explore states with different total energy, while the "dynamic" moves are achieved by using the Hamiltonian dynamics, allowing the algorithm to explore states with the total energy approximately constant. In simple form, the hybrid Monte Carlo method can be viewed as a combination of the Monte Carlo sampling method, that is guided by the gradient of the probability distribution function at each state.

9.2.1 Stochastic Dynamics Model

As mentioned before, in statistical mechanics the positions and the momentum of all molecules at a given time in a physical system define the state space of the system at that time. The positions of the molecules define the potential energy of a system and the momentum defines the kinetic energy of the system. In this chapter, what is referred to in statistical mechanics as the *canonical distribution of the "potential energy"* is the posterior distribution in Equation 9.4. The canonical distribution of the system's kinetic energy is:

$$P(\{p\}) = \frac{1}{Z_K}\exp(-K(\{p\})) \\ = (2\pi)^{-n/2}\exp(-\frac{1}{2}\sum_i p_i^2) \tag{9.6}$$

In molecular dynamics p_i is the momentum of the *i*th molecule. Here, p is not to be confused with P, which indicates probability. In neural networks, p_i is a fictitious parameter that is used to give the procedure a molecular dynamic structure. It should be noted that the weight vector, $\{w\}$, and momentum vector, $\{p\}$, are of the same size, so it is for that reason that the superscript W was used in Equation 9.4. The combined kinetic and potential energy is called the *Hamiltonian* of the system and can be written as (Neal, 1994; Bishop, 1995; Marwala, 2001):

$$H(w,p) = \beta\sum^{N}\sum_{k}^{K}\{y_{nk} - t_{nk}\}^2 + \frac{\alpha}{2}\sum_{j=1}^{W} w_j^2 + \frac{1}{2}\sum_{i}^{W} p_i^2 \tag{9.7}$$

In Equation 9.7, the first two terms are the potential energy of the system, which is the exponent of the posterior distribution of Equation 9.7, and the last term is the kinetic energy. The canonical distribution over the phase space, *i.e.*, the position and momentum, can be written as (Neal, 1994; Bishop 1995):

$$P(w,p) = \frac{1}{Z}\exp(-H(w,p)) = P(w \mid D)P(p) \tag{9.8}$$

By sampling through the distribution in Equation 9.8, the posterior distribution of weights is obtained by ignoring the distribution of the momentum vector, p. The dynamics in the phase space may be specified in terms of the Hamiltonian dynamics by expressing the derivative of the "position" and "momentum" in terms of fictitious time, τ. It should be recalled here that the word "position" used here is synonymous to network weights. The dynamics of the system may thus be written by using the Hamiltonian dynamics as (Neal, 1994; Bishop 1995):

$$\frac{dw_i}{d\tau} = +\frac{\partial H}{\partial p_i} = p_i \tag{9.9}$$

$$\frac{dp_i}{d\tau} = +\frac{\partial H}{\partial w_i} = -\frac{\partial E}{\partial p_i} \tag{9.10}$$

The dynamics specified in Equations 9.9 and 9.10, cannot be followed exactly and, as a result, these equations are discretized using a "leapfrog" method. The leapfrog discretization of Equations 9.9 to 9.10 may be written as (Neal, 1994; Bishop 1995):

$$\hat{p}_i(\tau+\frac{\varepsilon}{2}) = \hat{p}_i(\tau) - \frac{\varepsilon}{2}\frac{\partial E}{\partial w_i}(\hat{w}(\tau)) \tag{9.11}$$

$$\hat{w}_i(\tau+\varepsilon) = \hat{w}_i(\tau) + \varepsilon\hat{p}_i(\tau+\frac{\varepsilon}{2}) \tag{9.12}$$

$$\hat{p}_i(\tau+\varepsilon) = \hat{p}_i(\tau+\frac{\varepsilon}{2}) - \frac{\varepsilon}{2}\frac{\partial E}{\partial w_i}(\hat{w}(\tau+\varepsilon)) \tag{9.13}$$

Equations 9.11 to 9.13 can be explained as follows:

- using Equation 9.11, the leapfrog takes a little half-step for the momentum vector, $\{p\}$;
- using Equation 9.12, it takes a full step for the "position", $\{w\}$; and
- using Equation 9.13, it takes a half-step for the momentum vector, $\{p\}$.

The combination of these three steps forms a single leapfrog iteration that calculates the "position" and "momentum" of a system at time $\tau + \varepsilon$ from the network weight vector and "momentum" at time τ. The above discretization has the following characteristics:

- it is reversible in time;
- it almost conserves the Hamiltonian, representing the total energy; and
- it preserves the volume in the phase space, as required by Liouville's theorem (Neal, 1993). Volume preservation is achieved because the moves the leapfrog steps take are shear transformations.

One issue that should be noted is that following Hamiltonian dynamics does not sample through the canonical distribution, as represented by Equation 9.4 ergodically because the total energy remains constant, but rather at most, samples through the micro-canonical distribution for a given energy. One technique used to ensure that the simulation is ergodic, is by introducing "stochastic" moves by changing the Hamiltonian, H, during the simulation – achieved by replacing the "momentum" vector, $\{p\}$, before the next leapfrog iteration is performed. In this chapter, a normally distributed vector with a zero-mean replaces the "momentum" vector.

The dynamic steps introduced in this section make use of the gradient of the error with respect to the "position" (network weights) as shown in Equation 9.11. In this section, a procedure on how to move from one state to another is described. This procedure uses Hamiltonian dynamics to achieve dynamic moves and randomly changes the "momentum" vector to achieve stochastic moves. The next section describes how the states visited are either accepted or rejected.

9.2.2 Metropolis Algorithm

An algorithm due to Metropolis *et al.* (1953) has been used extensively to solve problems of statistical mechanics. In the Metropolis algorithm, on sampling a stochastic process $\{X_1,X_2,\ldots, X_n\}$ consisting of random variables, random changes to X are considered and are either accepted or rejected according to the following criterion:

$$\begin{aligned}&if\ H_{new} < H_{old}\ accept\ \ state\ \ (w_{new}, p_{new})\\&else\\&accept\ \ (w_{new}, p_{new})\ \ with\ \ probabilit\ y\\&\exp\{-(H_{new} - H_{old})\}\end{aligned} \tag{9.14}$$

In this chapter, this procedure is viewed as a way of generating a Markov chain with the transition from one state to another conducted using the criterion in Equation 9.14. By investigating Equation 9.14 carefully, it is observed that states with a high probability form the majority of the Markov chain, and those with low probability form the minority of the Markov chain. However, simulating a distribution by perturbing a single vector, $\{w\}$ is infeasible due to the high-dimensional nature of the state space and the variation of the posterior probability of weight vector. A technique that exploits the gradient of the Hamiltonian with respect to the weight vector $\{w\}$ was used to improve the Metropolis algorithm and is the subject of the next section.

9.2.3 Hybrid Monte Carlo

A Hybrid Monte Carlo combines a stochastic dynamics model with the Metropolis algorithm and, by so doing, eliminates the bias introduced by using a nonzero step size, as shown in Equations 9.11 to 9.13. The hybrid Monte Carlo method works by taking a series of trajectories from an initial state, *i.e.*, "positions" and "momentum", and moving in some direction in the state space for a given length of time and accepting the final state using the Metropolis algorithm. The validity of the hybrid Monte Carlo rests on three properties of the Hamiltonian dynamics:

- time reversibility: it is invariant under $t \to -t$, $p \to -p$;
- conservation of energy: the $H(w, p)$ is the same at all times; and

- conservation of state-space volumes due to Liouville's theorem (Neal, 1993).

For a given leapfrog step size, ε_0, and the number of leapfrog steps, L, the dynamic transition of the hybrid Monte Carlo procedure was conducted as follows:

1. randomly choose the direction of the trajectory, λ, to be either –1 for the backward trajectory or +1 for the forward trajectory;
2. starting from the initial state, $(\{w\},\{p\})$, perform L leapfrog steps (Equations 9.11 to 9.13) with the step size, $\varepsilon = \varepsilon_0(1 + 0.1k)$ resulting in state $(\{w\}^*,\{p\}^*)$, (Here, ε_0 is a chosen fixed step size and k is the number chosen from a uniform distribution and lies between 0 and 1;
3. reject or accept $(\{w\}^*,\{p\}^*)$ using the Metropolis criterion;
4. if the state is accepted then the new state becomes $(\{w\}^*,\{p\}^*)$; and
5. if rejected, the old state, $(\{w\}, \{p\})$, is retained as a new state.

After implementing Step 3 the momentum vector was reinitialized before moving on to generate the subsequent state. In this chapter, the momentum vector was sampled from a Gaussian distribution before generating the subsequent state. This ensured that the stochastic dynamics model samples were not restricted to the micro-canonical ensemble. By replacing the momenta, the total energy was allowed to vary because the momenta of the particles were refreshed. This idea of replacing the momentum was introduced by Anderson (1980).

One remark that should be noted about the hybrid Monte Carlo method is that it makes use of the gradient information in Step 2 above, using the leapfrog steps in Equation 9.19. The advantage of using this gradient information is that the hybrid Monte Carlo trajectories move in the direction of high probabilities, resulting in an improved probability that the resulting state is accepted and that the accepted states are not highly correlated. In neural networks, the gradient is calculated using back-propagation (Bishop, 1995).

The number of leapfrog steps, L, must be significantly higher than one to allow for a faster exploration of the state space. The choice of ε_0 and L affects the speed at which the simulation converges to a stationary distribution and the correlation between the states accepted. The leapfrog discretization does not introduce systematic errors due to occasional rejection of states that result in the increase of the Hamiltonian.

In Step 2 of the implementation of the hybrid Monte Carlo method, the step size $\varepsilon = \varepsilon_0(1 + 0.1k)$, where k is uniformly distributed between 0 and 1, is not fixed. In effect, this ensures that the actual step size for each trajectory is varied so that the accepted states do not have a high correlation (Mackenzie, 1989). The same effect can be achieved by varying the leapfrog steps. In this chapter, only the step size is varied.

The application of the Bayesian approach to neural networks results in weight vectors that have a mean and standard deviation and thus have a probability distribution. As a result, the output parameters have a probability distribution. Following the rules of probability theory, the distribution of the output vector $\{y\}$ for a given input vector $\{x\}$ may be written in the following form:

$$p(\{y\}|\{x\},D) = \int p(\{y\}|\{x\},\{w\})p(\{w\}|D)d\{w\} \tag{9.15}$$

In this chapter, the hybrid Monte Carlo method was employed to determine the distribution of the weight vectors and, subsequently, of the output parameters. The integral in Equation 9.15 may be approximated as follows:

$$I \equiv \frac{1}{L}\sum_{i=1}^{L} f(\{w\}_i) \tag{9.16}$$

In Equation 9.16, L is the number of retained states and f is the multi-layer perceptron network.

The application of a Bayesian framework to neural networks results in a mapping weight vector between the input and output with a probability distribution. This is different from the network implemented where the mapping weight vector did not have a probability distribution. In Figure 9.1, the multi-layer perceptron mapping function has a weight vector that forms a probability distribution function. These weight vectors are L in number, and come from the retained states resulting from a hybrid Monte Carlo sampling process.

9.3 Finite-element-model Updating Using Neural Networks and Control Theory

In this chapter, a Bayesian neural-network-based control approach was used to update a finite-element model. The reference signal was viewed as the measured data and the control process was aimed at identifying a set of points, *i.e.,* updated parameters that ensure the realization of the reference signal, *i.e.,* measured data. Xie *et al.* (2009) applied fuzzy control and neural networks for controlling a spark-ignition system while Kuo *et al.* (2007) applied a recurrent fuzzy neural network in controlling a two-phase linear brushless machine control.

Melin and Castillo (2007) successfully applied a hybrid of neural networks, fuzzy logic and fractal techniques for industrial quality control, while Villalva and Filho (2006) applied neural-network techniques for controlling a shunt active power filter.

Gao *et al.* (2004) successfully applied neural-network modeling for control of the grasping and manipulation of a hand-held object, while Shayeghi and Shayanfar (2004) applied radial basis functions for controlling a power-system load frequency. Other successful applications of the control system include:

- cement mills (Topalov and Kaynak, 2004);
- vehicle suspension (Cui *et al.*, 2004); and
- identifying neural-network architecture (Yang *et al.*, 2004).

In this chapter, the control framework represented in Figure 9.1 was applied for finite-element-model updating. In this chapter, the finite-element-model updating process is conducted by minimizing the following equation:

$$E = \sum \left(h(w, X) - Y_{ref}\right)^2 \tag{9.17}$$

Here, $h(w, X)$ is the neural-network model that is trained as explained earlier in the chapter; w is the network weights; X is the updating parameters; and Y_{ref} is the desired natural frequencies.

In Figure 9.1, the following procedure is illustrated:

1. given Y_{ref} as the desired natural frequencies from measurements;
2. initialize the updating parameters by defining the initial updating parameters X_0;
3. use the hybrid of particle-swarm optimization and the Nelder–Mead simplex method to minimize Equation 9.17 so as to identify the updated finite-element parameters X_0;
4. input the updated parameters into the finite-element model and then calculate the natural frequencies;
5. compare the predicted results from the finite-element model and measurements and if the differences are sufficiently large, repeat Step 3 using the updated parameters identified in Step 4; and
6. otherwise terminate.

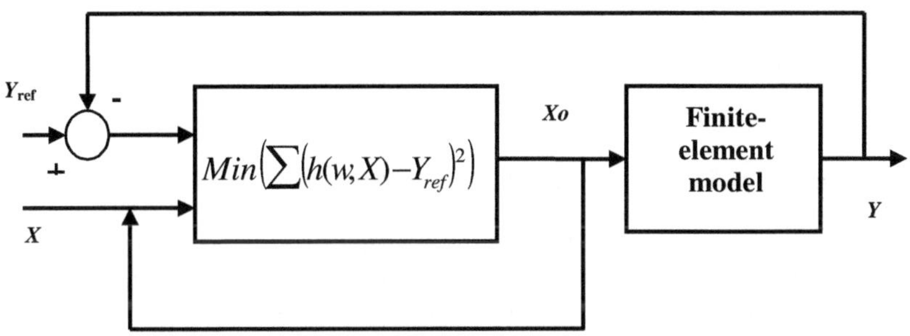

Figure 9.1 Control-based finite-element-model updating procedure

The hybrid of particle-swarm optimization and Nelder–Mead simplex method was used for Step 3 above. The hybrid procedure used in this chapter was explained in Chapter 7 and was motivated by the need to create an optimization method that has two core characteristics:

- having a global search capability normally found in an evolutionary type of optimization such as particle-swarm optimization as explained in Chapter 4; and

- having a local search capability that is intended to prevent premature convergence on a solution that is not necessarily optimal, such as the Nelder–Mead simplex method as explained in Chapter 2.

The procedure presented in this chapter was compared to the response-surface method studied in Chapter 6. Both these methods use neural networks. As explained in Chapter 6, the response-surface method is a procedure that operates by generating a response for a given input. In Chapter 6, a multi-layer perceptron neural network was used to construct the response equation.

In this chapter, the inputs were the parameters to be updated and the response was the error between the measured data and the finite-element-model-generated data. An approximation model of the input parameters and the response, called a *response-surface equation*, was then constructed. As a consequence of this, the optimization method and thus the finite-element-model updating process operate on the surface response. The main differences between the method presented in this chapter and the one in Chapter 6 is that neural networks were used to map the relationship between the updating parameters and the measured data in this chapter, while it was used to map the relationship between the updating parameters and the error in Chapter 6.

9.4 Example 1: A Simple Beam

The aluminum beam shown in Chapter 2 was used to test the presented approach. This beam had the following dimensions:

- length: 1.1 m;
- width: 29.2 mm; and
- thickness: 9.6 mm.

This beam was pierced with holes of diameter 5.8 mm, located at the centers of 8 elements and was therefore difficult to model. Further details of this beam were reported in Marwala (1997). The beam was tested freely suspended using elastic rubber bands. The beam was excited using an electromagnetic shaker and the response was measured using an accelerometer. The beam was also modeled using the structural dynamics toolbox (Balmès, 1997) and the beam was divided into 11 elements. The finite-element model used Euler–Bernoulli beam elements. It was excited at various positions and the acceleration was measured at 10 different positions.

A set of 10 frequency-response functions were calculated and a roving accelerometer was used for the testing. The moduli of elasticity of these elements were used as updating parameters. In implementing the finite-element-model updating, the moduli of elasticity were restricted to vary from 6×10^{10} to 8×10^{10} N m^{-2}.

The MATLAB® and NETLAB® toolbox (Nabney, 2001) were used to implement all the neural networks. The Bayesian network had an 11-input node, 9-hidden unit, and 4-output node (11–9–4) structure and was trained using hybrid Monte Carlo simulation by returning 1000 samples to form the posterior probability. On implementing the presented particle-swarm optimization method for finite-element-model updating the following parameters were chosen: the population was 50; $c_1 = 0.05$; $c_2 = 0.01$; and $w = 0.002$.

The particle-swarm-optimization method was run for 200 generations. The results of the presented method were compared to the results from the response-surface method (RSM) in Chapter 6. These results are summarized in Table 9.1.

When a modulus of elasticity of 7×10^{10} N m^{-2} was assumed, the error between the first measured natural frequency and that from the initial finite-element model was 1.9%. When the presented Bayesian neural-network (BNN)-based method was used for finite-element-model updating, this error was reduced to 0.5%, but using the RSM reduced it to 1.7%.

The error between the second measured natural frequency and that from the initial model was 2.2%. When the BNN method was used, this error was reduced to 1.2% but by using the RSM it was reduced to 0.7%.

The error of the third natural frequencies between the measured data and the initial finite-element model was 1.2%. When the BNN method was used, this error was increased to 0.8% and using the RSM it was reduced to 1.0%.

The error between the fourth measured natural frequency and that from the initial model was 1.4%. When the BNN method was used, this error was reduced to 0.2% and on using the RSM it was reduced to 0.4%.

Table 9.1 Results showing measured frequencies of the beam, the initial frequencies and the frequencies obtained when the finite-element model of a beam was updated using the Bayesian neural-network (BNN)-based updated model and the response-surface method (RSM)

Modes	Measured frequency (Hz)	Initial frequency (Hz)	Frequencies from the Bayesian neural-network-based updated Model (Hz)	Frequencies from the RSM updated model (Hz)
1	41.5	42.3	41.3	40.8
2	114.5	117.0	113.1	113.7
3	224.5	227.3	226.2	222.1
4	371.6	376.9	372.3	370.2

Overall, the Bayesian neural-network-based method gives the best results with an average error of 0.7%, while the RSM gives an average error of 1.0%. On average, both methods improved when compared to the average error between the initial finite-element model and the measured data.

The updated models implemented were also validated on the mode shapes they predicted. To make this assessment possible, the modal assurance criterion (MAC) was used (Allemang and Brown, 1982), and the results are shown in Table 9.2. The mean of the diagonal of the MAC vector was used to compare the mode shapes predicted by the updated and initial finite-element models with the measured mode shapes. An average value of 1.0 indicates that the mode shapes are perfectly correlated. The average MAC calculated between the mode shapes from an initial finite-element model and the measured mode shapes was 0.9986. When the average MAC was calculated between the measured data and data obtained from the updated finite-element models, it was observed that both models gave marginally improved averages for the diagonals of the MAC matrices of 0.9988.

Table 9.2 Beam results showing the MAC calculated between measured mode shapes and the initial finite-element model, the RSM updated finite-element model and the BNN updated finite-element model

Method	MAC
Initial model	0.9986
RSM	0.9988
BNN	0.9988

9.5 Example 2: Unsymmetrical H-shaped Structure

The unsymmetrical H-shaped aluminum structure shown in Chapter 2 was also used to validate the proposed method. This structure was also used by Marwala and Hunt (1999) and by Marwala (1997). The structure was excited using an electromagnetic shaker and the response was measured using an accelerometer. The structure was divided into 12 elements. It was excited and the acceleration was measured at 15 positions. The structure was tested freely suspended, and a set of 15 frequency-response functions were calculated. A roving accelerometer was used in the testing. The mass of the accelerometer was found to be negligible compared to the mass of the structure. As in the previous example, the finite-element model was constructed using the structural dynamics toolbox (Balmès, 1997) with the Euler–Bernoulli beam elements. The finite-element model contained 12 elements. As in the previous example, the moduli of elasticity of these elements were used as updating parameters, which were restricted to fall in the interval 6×10^{10} to 8×10^{10} N m^{-2}. The BNN method and RSM were implemented as in the previous example.

The results obtained when the BNN method and RSM were used for finite-element-model updating are shown in Table 9.3. Table 9.3 shows the measured natural frequencies; initial natural frequencies and natural frequencies obtained by the BNN method and genetic RSM-updated finite-element models.

Table 9.3 Results from an unsymmetrical H-shaped structure showing measured frequencies, the initial frequencies and the frequencies obtained when the finite-element model was updated using the BNN and RSM

Modes	Measured frequency (Hz)	Initial frequency (Hz)	Frequencies from the BNN updated model (Hz)	Frequencies from the RSM updated model (Hz)
1	59.9	56.2	51.3	52.0
2	117.3	127.1	117.8	118.7
3	208.4	228.4	206.7	209.9
4	254.0	269.4	255.6	251.6
5	445.1	452.4	451.5	432.5

The MATLAB® and NETLAB® toolbox (Nabney, 2007) were used to implement all the neural networks. The Bayesian network had a 12-input node, 9-hidden unit and 5-output node (11–9–11) structure and was trained using a hybrid Monte Carlo simulation to return 1000 samples to form the posterior probability. On implementing the presented particle-swarm-optimization method for finite-element-model updating, a population of 50, $c_1 = 0.05$ and $c_2 = 0.01$; and $w = 0.002$, were used. The particle-swarm-optimization method was run for 200 generations. The results of the presented method were compared to the results of the response-surface method of Chapter 6.

When a modulus of elasticity of 7×10^{10} N m^{-2} was assumed, the error between the first measured natural frequency and that from the initial finite-element model was 4.3%. When the BNN method was used for finite-element-model updating, this error was reduced to 4.8%, and the RSM approach reduced this error to 3.5%.

The error between the second measured natural frequency and that from the initial model was 8.4%. When the BNN method was used, the error was reduced to 0.4% and the RSM reduced this error to 1.2%.

The error of the third natural frequencies between the measured data and the initial finite-element model was 9.5%. When the BNN method was used, this error was reduced to 0.8% and using the RSM it was reduced to 0.7%.

The error between the fourth measured natural frequency and that from the initial model was 3.7%. When the BNN method was used, this error was reduced to 0.6% and using the RSM it was reduced to 0.9%.

The error between the fifth measured natural frequency and that of the initial model was 1.6%. When the BNN method was used, this error was increased to 1.4% and the RSM increased it to 2.8%. Overall, the BNN method gave the best results.

As in the previous example, the updated models implemented were validated on the mode shapes they predicted using the MAC. The results are in Table 9.4. Table 9.4 shows that the BNN and RSM updated finite-element models gave the improved averages of the diagonals of the MAC matrices of 0.8410 and 0.8413, respectively.

Table 9.4 Results of the unsymmetrical H-shaped structure showing measured frequencies, the initial frequencies and the frequencies obtained when the finite-element model was updated using the RSM and BNN

Method	MAC
Initial model	0.8394
RSM	0.8413
BNN	0.8410

9.6 Conclusion

In this study, the BNN method and the RSM were implemented for finite-element-model updating. When these techniques were tested on a simple beam and an unsymmetrical H-shaped structure, it was observed on average that the BNN gave more accurate results than the RSM method.

9.7 Future Work

This chapter introduced Bayesian neural networks for finite-element-model updating. For further work, other learning methods such as support vector machines should be used. The conclusions reached in this chapter are highly dependent on the nature of the data used in the analysis. Therefore, further statistical tests should be conducted to ensure that the conclusions reached are not dependent on the data used. In particular, the relationship between the nature of data and the finite-element-model updating process to be used must be identified.

References

Ait-Salem Duque O, Senin AR, Stenti A, De Munck M, Aparicio F (2007) A Methodology for the Choice of the Initial Conditions in the Model Updating of Welded Joints Using the Fuzzy Finite Element Method. Comput and Struct 85:1534–1546

Allemang RJ, Brown DL (1982) A Correlation Co-efficient for Modal Vector Analysis. In:Proc of the 1st Intl Modal Analysis Conf:01–18

Alpaydın E (2004) Introduction to Machine Learning (Adaptive Computation and Machine Learning). MIT Press, Cambridge

Anderson HC (1980) Molecular Dynamics Simulations at Constant Pressure and/or Temperature. J of Chem Phys 72:2384–2393

Atalla MJ (1999) Model Updating Using Neural Networks. PhD Thesis, Virginia Polytechnic Institute and State University

Balmès E (1997) Structural Dynamics Toolbox User's Manual. Sci Softw Group Sèvres Fr Ver 2.1

Bhagat PM (2005) Pattern Recognition in Industry. Elsevier, Amsterdam

Bishop CM (1995) Neural Networks for Pattern Recognition. Oxford University Press, London

Bogaerts A (2009) Hybrid Monte Carlo - Fluid Model for Studying the Effects of Nitrogen Addition to Argon Glow Discharges. Spectrochim Acta–Part B At Spectrosc 64:126–140

Chang CC, Chang TYP, Xu YG (2000) Adaptive Neural Networks for Model Updating of Structures. Smart Mater and Struct 9:59–68

Chang CC, Chang TYP, Xu YG, To WM (2002) Selection of Training Samples for Model Updating Using Neural Networks. J of Sound and Vib 249:867–883

Cheung SH, Beck JL (2009) Bayesian Model Updating Using Hybrid Monte Carlo Simulation with Application to Structural Dynamic Models with Many Uncertain Parameters. J of Eng Mech 135:243–255

Cui X, Chen L, Li D (2004) Vehicle-Semi-Active Suspension Adaptive Fuzzy Control Based on Neural Network Control Theory. China Mech Eng 15:178

Desgranges C, Delhommelle J (2008) Crystallization Mechanisms for Supercooled Liquid Xe at High Pressure and Temperature: Hybrid Monte Carlo Molecular Simulations. Phys Rev B - Condens Matter and Mater Phys 77:054201

Desgranges C, Delhommelle J (2009) Phase Equilibria of Molecular Fluids via Hybrid Monte Carlo Wang-Landau Simulations: Applications to Benzene and n-alkanes. J of Chem Phys 130:244109

Fei QG, Li AQ, Zhang LM (2005) Study on Finite Element Model Updating of Nonlinear Structures Using Neural Network. J of Astronaut 26:267–269

Fei QG, Zhang LM (2004) Finite Element Model Updating Using Radial Basis Function Neural Network. J of Nanjing, Univ of Aeronaut and Astronaut 36:748–752

Friswell MI, Mottershead JE (1995) Finite Element Model Updating in Structural Dynamics. Kluwer Academic Publishers Group, Norwell

Gao F, Latash ML, Zatsiorsky VM (2004) Neural Network Modeling Supports a Theory on the Hierarchical Control of Prehension. Neural Comput and Appl 13:352–359

Gnewuch M (2009) On Probabilistic Results for the Discrepancy of a Hybrid-Monte Carlo Sequence. J of Complex 25:312–317

Haykin S (1999) Neural Networks. Prentice-Hall, New Jersey

He HX, Yan WM, Wang Z (2008) Stepwise Model Updating Method Based on Substructures and GA-ANN. Eng Mech 25:99–105

Henelius P, Fishman RS (2008) Hybrid Quantum-Classical Monte Carlo Study of a Molecule-Based Magnet. Phys Rev B–Condens Matter and Mater Phys 78:214405

Hurwitz E, Marwala T (2005) Optimising Reinforcement Learning for Neural Networks. In: Proc of the 6th Annu Eur on Intell Games and Simul, UK:13–18

Kułak L (2009) Hybrid Monte-Carlo Simulations of Fluorescence Anisotropy Decay in Three-Component Donor-Mediator-Acceptor Systems in the Presence of Energy Transfer. Chem Phys Lett 467:435–438

Kuo JL, Chang ZS, Lee JD (2007) Two-Phase Linear Brushless Machine Control Approach via Recurrent Fuzzy Neural Network Theory. WSEAS Trans on Syst 6:725–730

Leke BB, Marwala T, Tim T, Lagazio M (2006) Prediction of HIV Status from Demographic Data Using Neural Networks. In: Proc of the IEEE Intl Conf on Syst, Man and Cybern:2339–2344

Levin RI, Lieven NAJ (1998) Dynamic Finite Element Model Updating Using Neural Networks. J of Sound and Vib 210:593–607

Levin RI, Lieven NAJ, Lowenberg MH (2000) Measuring and Improving Neural Network Generalization for Model Updating. J of Sound and Vib 238:401–424

Liu Y, Duan Z, Li H (2009) Updating of Finite Element Model in Considering Mode Errors with Fuzzy Theory. Key Eng Mater 413–414:785–792

Lu Y, Tu Z (2004) A Two-Level Neural Network Approach for Dynamic FE Model Updating Including Damping. J of Sound and Vib 275:931–952

Machowski LA, Marwala T (2004) Representing and Matching 2D Shapes of Natural Objects Using Neural Networks. In: Proc of the IEEE Intl Conf on Syst, Man and Cybern:6366 6372

Mackenzie PB (1989) An Improved Hybrid Monte Carlo Method. Phys Lett B 226:369–371

Maiti S, Tiwari RK (2009) A Hybrid Monte Carlo Method Based Artificial Neural Networks Approach for Rock Boundaries Identification: A Case Study from the KTB Bore Hole. Pure and Appl Geophys 166:2059–2090

Marivate V, Ssali G, Marwala T (2008) An Intelligent Multi-Agent Recommender System for Human Capacity Building. In: Proc of the 14th IEEE Mediterr Electrotech Conf:909–915

Marwala T (1997) A Multiple Criterion Updating Method for Damage Detection on Structures. MSc Thesis, Univ of Pretoria

Marwala T (2001) Fault Identification Using Neural Networks and Vibration Data. PhD Thesis, Univ of Cambridge

Marwala T (2004) Control of Complex Systems Using Bayesian Neural Networks and Genetic Algorithm. Intl J of Eng Simul 5:28–37

Marwala T (2004) Fault Classification Using Pseudo Modal Energies and Probabilistic Neural Networks. Am Soc of Civ Eng, J of Eng Mech 130: 1346–1355

Marwala T (2007) Bayesian Training of Neural Network Using Genetic Programming. Pattern Recognit Lett 28:1452–1458

Marwala T, Hunt HEM (1999) Fault Identification Using Finite Element Models and Neural Networks. Mech Syst and Signal Process 13:475–490

Marwala T, Sibisi S (2005) Finite Element Updating Using Bayesian Framework and Modal Properties. Am Inst of Aeronaut and Astronaut, J of Aircr 42:275–278

Melin P, Castillo O (2007) An Intelligent Hybrid Approach for Industrial Quality Control Combining Neural Networks, Fuzzy Logic and Fractal Theory. Inf Sci 177:1543–1557

Metropolis N, Rosenbluth AW, Rosenbluth MN, Teller AH, Teller E (1953) Equations of State Calculations by Fast Computing Machines. J of Chem Phys 21:1087–1092

Mohamed N, Rubin D, Marwala T (2006) Detection of Epileptiform Activity in Human EEG Signals Using Bayesian Neural Networks. Neural Inf Process – Lett and Rev 10:01–10

Mohamed S, Marwala T, Rubin D (2007) Adaptive GPCR Classification Based on Incremental Learning. SAIEE Afr Res J 98:71–80

Msiza IS, Nelwamondo FV, Marwala T (2008) Water Demand Prediction Using Artificial Neural Networks and Support Vector Regression. J of Comput 3: 01–08

Mthembu L, Marwala T, Friswell MI, Adhikari S (2009) Bayesian Evidence for Finite Element Model Updating. In: Proc of the IMAC XXVII

Nabney IT (2001) Netlab: Algorithms for Pattern Recognition. Heidelberg: Springer-Verlag

Neal RM (1992) Bayesian Training of Backpropagation Networks by Hybrid Monte Carlo Method. Univ of Toronto Tech Rep CRG-TR-92-1

Neal RM (1993) Probabilistic Inference Using Markov Chain Monte Carlo Methods. Univ of Toronto Tech Rep CRG-TR-93-1

Neal RM (1994) Bayesian Learning for Neural Networks. PhD Thesis, Univ of Toronto, Canada

Nelwamondo FV, Mahola U, Marwala T (2006) Multi-Scale Fractal Dimension for Speaker Identification System. Trans on Syst 5:1152–1157

Patel P, Marwala T (2006) Neural Networks Fuzzy Inference Systems and Adaptive-Neuro Fuzzy Inference Systems for Financial Decision Making. Lect Notes in Comput Sci 4234:430–439

Perez M, Rubin DM, Marwala T, Scott LE, Stevens W (2008) A Hybrid Fuzzy-SVM Classifier Applied to Gene Expression Profiling for Automated Leukaemia Diagnosis. In: Proc of the IEEE Conf, Isr:041–045

Shayeghi H, Shayanfar HA (2004) Power System Load Frequency Control Using RBF Neural Networks Based on μ-Synthesis Theory. IEEE Conf on Cybern and Intell Syst:93–98

Takaishi T (2008) Financial Time Series Analysis of SV Model by Hybrid Monte Carlo. Lect Notes in Comput Sci (Incl Subseries Lect Notes in Artif Intell and Lect Notes in Bioinformatics) 5226:929–936

Tan D, Qu W, Wang J (2009) The Finite Element Model Updating of Structure Based on Wavelet Packet Analysis and Support Vector Machines. J of Huazhong Univ of Sci and Tech 37:104–107

Topalov AV, Kaynak O (2004) Neural Network Modeling and Control of Cement Mills Using a Variable Structure Systems Theory Based on-line Learning Mechanism. J of Process Control 14:581–589

Vilakazi CB, Marwala T (2006) Bushing Fault Detection and Diagnosis Using Extension Neural Network. In: Proc of the 10th IEEE Intl Conf on Intell Eng Syst:170–174

Villalva MG, Filho ER (2006) Control of a Shunt Active Power Filter with Neural Networks - Theory and Practical Results. IEEJ Trans on Ind Appl 126:946–953

Vivarelli F, Williams CKI (2001) Comparing Bayesian Neural Network Algorithms for Classifying Segmented Outdoor Images. Neural Netw 14:427–437

Weber M, Becker R, Durmaz VA, Köppen R (2008) Classical Hybrid Monte-Carlo Simulation of the Interconversion of Hexabromocyclododecane Stereoisomers. Mol Simul 34:727–736

Xie CQ, E JQ, Cheng ZM, Gong JK, Jiang SS, Yuan WH (2009) Soft-Measuring Model of Time Difference About Spark-Ignition Advanced Angle Based on Fuzzy Control and Neural Networks Theory. Chin Intern Combust Engine Eng 30:73–77

Yang HW, Zhan YQ, Shi GL, Qiao JW (2004) Improving the Architecture-Based Neural Network Model by Using Hierarchical Control Theory. J of Shanghai Jiaotong Univ 38:1369–1372

Zapico JL, Gonzlez-Buelga A, Gonzlez MP, Alonso RA (2008) Finite Element Model Updating of a Small Steel Frame Using Neural Networks. Smart Mater and Struct 17:045016

Zhu Y, Zhang L (2009) Finite Element Model Updating Based on least Squares Support Vector Machines. Lect Notes in Comput Sci (Incl Subseries Lect Notes in Artif Intell and Lect Notes in Bioinformatics

Chapter 10

Finite-element-model Updating Using a Bayesian Approach

Abstract. This chapter implements a Bayesian approach to finite-element-model updating. The Bayesian formulation was solved using the Markov chain Monte Carlo (MCMC) technique, as well as genetic programming based on the MCMC. These methods were tested on a simple beam and an unsymmetrical H-shaped structure. It was observed on average that the genetic programming MCMC performed better than the MCMC method.

Keywords: Bayesian approach, Metropolis algorithm, genetic programming, Markov chain Monte Carlo, likelihood probability distribution function, prior probability distribution function, crossover, mutation, reproduction

10.1 Introduction

Essentially, the maximum-likelihood formulation entails the process of formulating an objective function and then optimizing it (von Mises, 1939; Ellis, 1843; Kendall, 1949). This is simply interpreted as a way of identifying the most likely model and it is often termed the *frequentist approach*. Even though this framework has been applied successfully as observed in earlier chapters, it has shortcomings:

- it does not offer the user confidence intervals for the optimum solutions it gives;
- there is no philosophical explanation of the regularization terms that are used to control the complexity of the updated model; and
- it cannot handle inherent ill-conditioning or nonuniqueness in the finite-element-model-updating problem.

For this chapter, a Bayesian framework was adopted to address the shortcomings explained above. The Bayesian approach is a process that originated from Bayes' theorem where the probability distribution of the model, given the observed data, can be deduced from the probability distribution of the data, given the model, the probability distribution of the model and the probability distribution of the data. The Bayesian framework has been found to offer several advantages over maximum-likelihood techniques in areas closely mirroring finite-element-model-updating. These include (Neal, 1993; Beck and Katafygiotis, 1998; Katafygiotis *et al.*, 1998; Marwala, 2002):

- exact inferences and confidence intervals that do not depend on sample estimations can be obtained by using the Bayesian technique;
- the elimination of problematic parameters is theoretically easy;
- a stopping set of laws is immaterial for a Bayesian analysis;
- it permits model flexibility;
- it offers a philosophical explanation for the regularization parameters; and
- it does not over-fit the data.

However, the Bayesian approach also has the following disadvantages (Marwala, 2007a; Marwala, 2009):

- it necessitates the identification of a prior distribution for all flexible variables;
- it usually entails integrals of many dimensions; and
- very often, it requires the implementation of computationally expensive procedures such as the Monte Carlo simulation.

This chapter addresses the following issues:

- how prior information can be incorporated into the finite-element-model-updating problem (Mottershead and Friswell, 1995); and
- how the Bayesian framework can be applied to update finite-element models to match experimentally measured modal properties (*i.e.,* the natural frequencies and mode shapes) to modal properties calculated from the finite-element model.

Ching and Leu (2009) applied a Bayesian approach for approximating time-varying reliabilities of civil infrastructure facilities in a situation where failure data of the infrastructure system was not present and only condition-state data of its components was present. This technique assumed that degradation could be modeled as a Poisson process with an unidentified time-varying arrival rate and damage impact, and that the target system could be modeled as a fault-free model. A Bayesian algorithm was used to model large uncertainties and was successfully tested on a hydraulic spillway gate system.

Grieco and Hogarth (2009) applied a Bayesian approach to study the degree of confidence for easy and difficult tasks, while Berahman and Behnamfar (2009) applied finite-element-model data and a Bayesian approach to evaluate unknown demand model parameters for steel petroleum storage tanks. Liang (2009) improved the convergence of a stochastic approximation Monte Carlo by using smoothing techniques and successfully applied this to Bayesian model selection problems.

Watanabe and Nakamura (2009) proposed a fast incremental adaptation Bayesian method, based on a macroscopic time-evolution system that estimated an utterance-by-utterance update by estimating the posterior distributions and used this to achieve an online adaptation of large-vocabulary continuous speech recognition. The results obtained showed a significant improvement in word accuracy. The application of a Bayesian model selection technique aided in the concurrent online adaptation and detection of environmental changes.

In this chapter, a Markov chain Monte Carlo (MCMC) technique and a genetic-programming-based MCMC method were used to sample the probability of the updating parameters in the light of the measured modal properties, within the context of the Bayesian framework. This probability is known as the *posterior probability*. The Metropolis algorithm (Metropolis *et al.*, 1953) uses the probability as an acceptance criterion when sampling. The next section describes the mathematical approaches that are used in this chapter.

10.2 Mathematical Foundation

10.2.1 Dynamics

All elastic structures may be described in terms of their distributed mass, damping, and stiffness matrices. If damping terms are neglected, the dynamic equation may be written in the modal domain (natural frequencies and mode shapes) for the *i*th mode as (Ewins, 1995):

$$(-\omega_i^2[M]+[K])\{\phi\}_i=\{\varepsilon\}_i \tag{10.1}$$

Here, $[M]$ is the mass matrix, $[K]$ is the stiffness matrix, ω_i is the *i*th natural frequency, $\{\phi\}_i$ is the *i*th mode shape vector and $\{\varepsilon\}_i$ is the *i*th error vector. The error vector $\{\varepsilon\}_i$ is equal to $\{0\}$ if the system matrices $[M]$ and $[K]$ correspond to the modal properties. If the system matrices, usually obtained from the finite-element model, do not match the measured modal properties ω_i and $\{\phi\}_i$ then $\{\varepsilon\}_i$ is a non-zero vector. In the maximum-likelihood method, the Euclidean norm of $\{\varepsilon\}_i$ is minimized in order to match the system matrices to the measured modal properties.

Another problem encountered in many practical situations is that the dimension of mode shapes does not match the dimension of the system matrices. This is because the measured modal coordinates are less than finite-element modal coordinates. To ensure compatibility between system matrices and mode shape vectors, the dimension of system matrices was reduced using a technique known as the *Guyan reduction method* (Guyan, 1965). The Guyan reduction method was described in Chapter 2 of this book, and was used here because of its simplicity of implementation while giving good results.

10.2.2 Bayesian Method

The Bayesian approach is a method that is based on Bayes' theorem. This theorem states that given the observed data, the probability of the model, also called *the posterior probability function,* depends on the likelihood function, the prior distribution, and the evidence (Hald, 1998; Stigler, 1990). Wu and Li (2004) applied a Bayesian estimation method for the identification of structural parameters and damage detection in a steel structure. This procedure used a finite-element-model updating technique, and the Bayesian estimation technique to identify the connection stiffness of beam-column joints and Young's modulus of the structure.

Reda Taha and Lucero (2004) applied a Bayesian technique to distinguish levels of damage into fuzzy sets, taking into account the uncertainty related to the confusing damage types. Their method was used together with finite-element models in a concrete bridge. Wu and Li (2004) applied finite-element-model updating of a tall structure to match ambient data better. The Bayesian approach was used to estimate the most appropriate and efficient technique for the finite-element-model updating.

Yuen and Katafygiotis (2001) applied the Bayesian method to finite-element-model updating using time-domain ambient data. Their results demonstrated that the updated probability density functions could be estimated by using a Gaussian distribution that was centered at the optimal parameters at which the updated probability density function was maximized.

Hongxing *et al.* (2000) successfully applied a Bayesian estimator as a multi-objective, multi-design-variable optimization method. In this chapter, the Bayesian method was introduced to solve the finite-element-updating problem based on modal properties. The fundamental rule that governs the Bayesian approach is written as follows (Bishop, 2006):

$$P(\{E\}|[D]) = \frac{P([D]|\{E\})P(\{E\})}{P([D])} \tag{10.2}$$

Here, $\{E\}$ is a vector of updating parameters; $P(\{E\})$ is the probability distribution function of updating parameters in the absence of any data, which is known as the *prior distribution;* and $[D]$ is the matrix that contains natural frequencies ω_i and mode shapes $\{\phi\}_i$.

It must be noted that the mass $[M]$ and stiffness $[K]$ matrices are functions of the updating parameter {E}. The quantity $P(\{E\}|[D])$ is the posterior probability distribution function after a set of data has been seen, $P([D]|\{E\})$ is the likelihood probability distribution function and $P([D])$ is a normalization factor.

A) Likelihood-probability Distribution Function

There are many areas where the likelihood-probability distribution function has been applied. These include neural networks (Bishop, 2006). In the neural-network context, the likelihood distribution function is defined as the normalized exponent of the error function. In this chapter, the likelihood distribution function,

$P([D]|\{E\})$, was defined as the sum of squares of elements of the error vector shown in Equation 10.1, and can be written in the same way as for neural networks as:

$$\begin{aligned} P([D]|\{E\}) &= \frac{1}{Z_D(\beta)}\exp\left(-\beta\sum_j^F\sum_i^N\left(\varepsilon_{ij}^2\right)\right) \\ &= \frac{1}{Z_D(\beta)}\exp\left(-\beta\sum_j^F\sum_i^N\left(\left[(-\omega_i^2[M]+[K])\{\phi\}_i\right]_j\right)^2\right) \end{aligned} \tag{10.3}$$

Here, β is the coefficient of the measured modal property data contribution to the error and is set to 1 through trial and error, and ε_{ij} is the error matrix with subscript i representing the ith modal properties and j represents the jth measurement position. The superscript F is the number of measured mode shape coordinates, N is the number of measured modes and Z_D is:

$$Z_D(\beta) = \int \exp\left(-\beta\sum_j^F\sum_i^N\left(\left[(-\omega_i^2[M]+[K])\{\phi\}_i\right]_j\right)^2\right)d[D] \tag{10.4}$$

It should be noted that, in Equation 10.3, the error ε_{ij} is a matrix as opposed to being a vector, as was the case in Equation 10.1. This is because it takes into account all the modal co-ordinates.

B) Prior probability Distribution Function of Parameters to Be Updated

The prior probability distribution function consists of the information that is known about the problem. In this chapter, it is known that not all parameters to be updated have the same level of modeling errors. This means that some parameters are to be updated more intensely than others. For example, parameters next to joints should be updated more intensely than for those having smooth surface areas far from joints. In this chapter, the prior probability distribution function for parameters to be updated may be written by using the Gaussian assumption as (Bishop, 1995):

$$P(\{E\}) = \frac{1}{Z_E(\alpha)}\exp\left(-\sum_i^Q\frac{\alpha_i}{2}\|\{E\}\|^2\right) \tag{10.5}$$

where Q is the number of groups of parameters to be updated, and α_i is the coefficient of the prior probability distribution function for the ith group of updating parameters.

The prior probability distribution function in Equation 10.5 ensures that large updating of parameters is less likely than small adjustments of updating parameters. The Gaussian prior has been successfully used to identify a large number of weights in neural networks, and therefore it was assumed that it should be successful on identifying a small number of updating parameters in this chapter. The higher α_i is, the lower the degree of updating the *i*th group of parameters. Furthermore, ||•|| is the Euclidean norm of •.

In Equation 10.5, if α_i is constant for all the updating parameters, then the updated parameters will be of the same order of magnitude. Equation 10.5 may be viewed as a regularization parameter (Vapnik, 1995). In Equation 10.5, Gaussian priors were conveniently chosen because many natural processes tend to have a Gaussian distribution. In a Bayesian framework, the regularization method is viewed as a mechanism of incorporating prior information, whereas in a maximum-likelihood method it is viewed as mathematical convenience. The function $Z_E(\alpha)$ is a normalization factor given by (Bishop, 2006):

$$Z_E(\alpha) = \int \exp\left(-\sum_i^Q \frac{\alpha_i}{2}\|\{E\}\|^2\right) d\alpha_i \tag{10.6}$$

C) Posterior Distribution Function of Updating Parameters Vector

The distribution of the updating parameters $P(\{E\}|[D])$ after the data have been seen is calculated by substituting Equations 10.3 and 10.5 into Equation 10.2 to give:

$$P(\{E\}|[D]) = \frac{1}{Z_s(\alpha,\beta)} \exp\left(-\beta \sum_j^F \sum_i^N \left(\left[-\omega_i^2 [M] + [K]\right)\{\phi\}_i\right]_j\right)^2 - \sum_i^Q \frac{\alpha_i}{2}\|\{E\}\|^2\right) \tag{10.7}$$

where,

$$\begin{aligned} Z_S(\alpha,\beta) &= P([D]) \\ &= \int \exp\left(-\beta \sum_j^F \sum_i^N \left(\left[(-\omega_i^2 [M] + [K])\{\phi\}_i\right]_j\right)^2 - \sum_i^Q \frac{\alpha_i}{2}\|\{E\}\|^2\right) d\{E\} \end{aligned} \tag{10.8}$$

In Equation 10.7, the optimal weight vector corresponds to the maximum of the posterior probability distribution function, which is the solution that is obtained from a maximum-likelihood approach. This implies that a Bayesian method gives at least the solution that is given by the maximum-likelihood method, and in addition it gives probability distributions.

10.2.3 Markov Chain Monte Carlo Method

The application of the Bayesian approach to finite-element-model updating using a Monte Carlo technique results in a set of updated parameter vectors $\{E\}_i$. Each vector can be used in conjunction with the finite-element model to predict the modal properties. Because there are many such vectors, they can be used to predict as many modal properties as there are updating vectors $\{E\}_i$. In fact, the distributions of these predicted modal properties may be constructed and their averages and variances may also be calculated. As a result, the dynamic predictions of natural frequencies and mode shapes will have probability distributions. Following the rules of probability theory, the distribution of vector $\{y\}$, representing measured modal properties may be written in the form:

$$P(\{y\}|[D]) = \int P(\{y\}|[E])P(\{E\}|[D])d\{E\} \tag{10.9}$$

Equation 10.9 depends on Equation 10.7, and is difficult to solve analytically due to the relatively high dimension of the updating parameter vector. As a result, a Markov chain Monte Carlo (MCMC) method was employed to determine the distribution of updating parameters and, subsequently, of predicted modal properties.

MCMC techniques are a group of methods that are used for sampling from probability distributions. These methods are based on building a Markov chain that has the required distribution as its equilibrium distribution. The state of the chain, following many steps, is then used as a sample from the required distribution and the quality of the sample gets better as the number of steps is increased (Gill, 2008).

Typically, it is easy to build a Markov chain with the required properties but the complicated issue is to decide on how many steps are necessary to converge to the stationary distribution with an acceptable error. Beginning from a random state, a high-quality chain rapidly achieves a stationary distribution.

The most regular use of the MCMC is to calculate integrals of many dimensions numerically (Green, 1995). Typically, an ensemble of "walkers" takes a random walk and, at each position where the walker arrives, the integrand value at that point is accumulated for the total integral (Neal, 2003).

The walker can subsequently construct a number of preliminary steps in the area, in search of a position to go after with high input into the integral. Random-walk techniques are an instance of the Monte Carlo approach. While the random samples of the integrand in a typical Monte Carlo integration are statistically self-governing, those used in the MCMC are correlated. A Markov chain is built in such way that the integrand becomes its equilibrium distribution and this is frequently simple to achieve (Robert and Casella, 2004).

Goodman and Lin (2009) controlled the coupling variates for MCMC to increase the accuracy of MCMC calculations in cases where the steady-state probability distribution is not clearly identified. Campillo *et al.* (2009) proposed a method for using M parallel MCMCs that work together with the intention of getting an estimate of an independent *M*-sample of a given intended rule and successfully applied this in a biomass evolution model.

Gallagher *et al.* (2009) applied MCMC sampling techniques to identify optimal models, model resolution, and model choice and applied this to an earth-science problem. Rudolf (2009) studied the explicit error bounds for a lazy reversible MCMC. Dimov *et al.* (2008) studied the robustness and practicality of MCMC methods for solving eigenvalue problems. Adequate conditions for building sensible Monte Carlo methods were deduced.

Tan (2008) applied the modified MCMC to build an estimator for computing integrals and expectations. The Markov chain method was viewed as a random design and an estimator was defined for the baseline measure. To achieve computational efficiency, methods such as subsampling, regulation, and amplification were used. The technique gave better results than the traditional Monte Carlo estimator.

Tonazzini and Bedini (2003) applied the MCMC method for unsupervised image denoising. Their method maximized the posterior distribution with respect to the line field and the Gibbs parameters. Their method was experimentally validated. Other applications of the MCMC include modeling epidemics (O'Neill, 2002).

Kosorok (2000) extended the Monte Carlo error-approximation techniques from univariate to multi-variate Markov chains, while Hum *et al.* (1999) proposed a block updating technique to tackle constrained MCMC sampling.

The integral in Equation 10.9 was solved, using an algorithm introduced by Metropolis *et al.* (1953), through generating a sequence of vectors $\{\{E\}_1,\{E\}_2,\ldots,\{E\}_n\}$ that form a Markov chain that converged to a stationary distribution $P([D]|\{E\})$. The integral in Equation 10.9 may thus be approximated as:

$$\left\{\tilde{y}\right\} \cong \frac{1}{L}\sum_{i=I}^{R+L-1} G(\{E\}_i) \tag{10.10}$$

Here, G is a finite-element model that takes vector $\{E\}_i$ and predicts the average output, $\{\tilde{y}\}$ is the vector containing the modal properties, R is the number of initial states that are discarded in the hope of reaching a stationary distribution described by Equation 10.7, and L is the number of retained states.

Several methods have been proposed to simulate the distribution in Equation 10.7 such as Gibbs sampling (Geman and Geman, 1984), the Metropolis algorithm and hybrid Monte Carlo methods (Duane *et al.*, 1987). The hybrid Monte Carlo, shown to be the most efficient of the Monte Carlo methods thus far, was not used in this chapter because it requires gradient information, which is not available in the exact form in the finite-element-model updating problem. This approach was, however, used in Chapter 9 to train neural networks. In this chapter the MCMC method is used to identify the posterior probability distribution function of the updating parameters.

The MCMC method was implemented by sampling a stochastic process consisting of random variables $\{\{E\}_1,\{E\}_2,\ldots,\{E\}_n\}$ through introducing random changes to the updating parameter vector $\{E\}$ and either accepting or rejecting the sample according to the Metropolis algorithm. The Metropolis criteria can be written as:

$$\begin{aligned}&if\ P_{new}(\{E\}|[D]) > P_{old}(\{E\}|[D])\ accept\ state\ \{E\}_{new}\\&else\ accept\ \{E\}_{new}\ with\ probabilit\ y\ \frac{P_{new}(\{E\}|[D])}{P_{old}(\{E\}|[D])}\end{aligned} \quad (10.11)$$

This chapter viewed this procedure as a way of generating a Markov chain with transition from one state to another conducted using the criterion in Equation 10.11. Marwala *et al.* (2007a) compared the maximum-likelihood method to the Bayesian method for finite-element-model updating. The maximum-likelihood method was implemented using a genetic algorithm, as in Chapter 3, while the Bayesian method was implemented using the MCMC. These methods were tested on a simple beam. The results showed that the Bayesian method gave updated finite-element models that more accurately predicted modal properties than the updated finite-element models obtained through the use of the maximum-likelihood method. Furthermore, both these methods were found to require the same computational load.

10.2.4 MCMC: Genetic Programming and Metropolis Algorithm

Genetic programming takes features of natural evolution and uses these to computationally solve practical problems. Genetic algorithms are examples of genetic programming and a procedure inspired by these will be introduced in this section. In this chapter, some of the features of genetic computing were applied to sample the posterior distribution function in Equation 10.3.

Marwala (2007b) applied genetic programming for Bayesian training of neural networks. The Bayesian neural networks were trained using the MCMC and genetic programming in binary space within the Metropolis framework. The algorithm could learn using samples obtained from previous steps and merged using concepts of natural evolution that included mutation, crossover, and reproduction. The reproduction function was the Metropolis framework. Binary mutation and simple crossover were used. The presented algorithm was tested on a simulated function, an artificial taster using measured data, as well as on condition monitoring of structures, and the results were compared to those from a classical MCMC method. The results confirmed that Bayesian neural networks trained using genetic programming offered better performance and efficiency than the classical approach.

Kouchakpour *et al.* (2009) developed the genetic programming method by dynamically varying the population size. It was observed that this method could give solutions with a reduced computational load over standard genetic programming.

Garcia-Arnau *et al.* (2006) proposed a tree-generation method for grammar-guided genetic programming that took into account a parameter to control the maximum size of the trees to be created. This procedure was then applied to the prognosis of breast cancer.

Tsakonas *et al.* (2006) used a genetic programming method for evolving rule-based systems in two medical domains, while Lones and Tyrrell (2004) applied genetic programming for modeling biological evolvability – particularly within the program representation used by enzyme genetic programming.

McKay (2001) examined the usage of partial functions, fitness sharing, and committee learning in genetic programming and applied this for learning spatial relationships for ecological modeling. Fonlupt (2001) applied genetic programming to solving the ocean's color. Other applications of genetic programming include:

- cancer research (Worzel *et al.*, 2009);
- designing classifiers (Lin *et al.*, 2007);
- designing intelligent structures (Tsakonas, 2006);
- optimal control (Kumar and Balasubramaniam, 2007);
- handwritten digit recognition (Parkins and Nandi, 2004); and
- trading (Oussaidene *et al.*, 1997).

Genetic algorithms were inspired by Darwin's theory of natural evolution. In natural evolution, members of the population compete with each other to survive and reproduce. Evolutionary successful individuals reproduce, while weaker members die. As a result, the successful genes are likely to spread within the population. This natural optimization method has been successfully used to optimize complex problems (Holland, 1975; Michalewicz, 1996; Goldberg, 1989). This procedure uses a population of binary-string chromosomes. Each of these strings is the discretized representation of a point in the search space and therefore has a fitness function that is given by the objective function. In generating a new population, three operators are performed:

- crossover;
- mutation; and
- reproduction.

These operators are adopted in genetic MCMC sampling. The crossover operator mixes genetic information in the population by cutting pairs of chromosomes at random points along their length and exchanging over the cut sections. This has a potential of joining successful operators together. Crossover occurs with a certain probability. In many natural systems, the probability of crossover occurring is higher than the probability of mutation occurring. A simple crossover technique (Goldberg, 1989) was used in this chapter. For a simple crossover, one crossover point was selected, a binary string from the beginning of the chromosome to the crossover point was copied from one parent, and the rest was copied from the second parent. For example, when 11001011 undergoes simple crossover with 11011111 it becomes 11001111.

The mutation operator picks a binary digit of the chromosomes at random and inverts it. This has a potential of introducing new information to the population. Mutation occurs with a certain probability. In many natural systems, the probability of mutation is low (*e.g.*, less than 1%). In this chapter, binary mutation (Goldberg, 1989) was used. When binary mutation is used, a number written in binary form is chosen and its value is inverted. For an example: 11001011 may become 11000011.

Reproduction takes successful chromosomes and reproduces them in accordance with their fitness functions. In this chapter, the Metropolis criterion was used as a reproduction method. By so doing, the least-fit members were therefore gradually driven out of the population of states that form a Markov chain.

A schematic illustration of the MCMC method trained using genetic programming is shown in Figure 10.1.

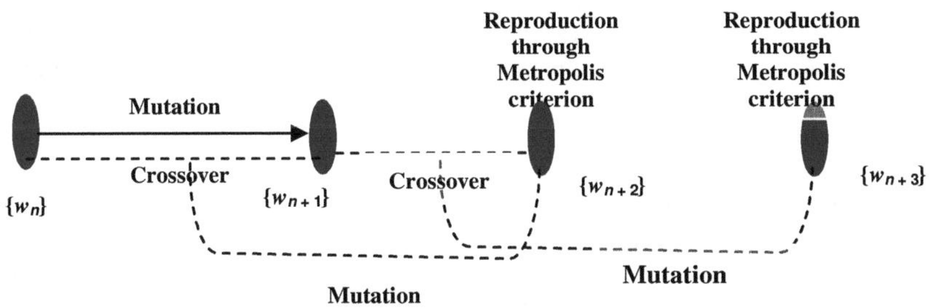

Figure 10.1 Schematic illustration of genetic sampling during the implementation of the MCMC

In this figure the following procedure is followed:

1. an initial sample updating parameter vector $\{E\}_n$ is generated;
2. then the sample is converted into binary form using the Gray code method (Michalewicz, 1996);
3. the sample is then mutated to form a new sample vector $\{E\}_{n+1}$;
4. the new updating parameters vector, $\{E\}_{n+1}$, undergoes crossover with its predecessor, $\{E\}_n$, and mutates again to form a new network updating parameters vector, $\{E\}_{n+2}$;
5. the updating parameters vector, $\{E\}_{n+2}$, is converted into floating-point and then its probability is calculated;
6. this updating parameters vector is accepted or rejected using the Metropolis criterion;
7. thereafter, States $\{E\}_{n+2}$ and $\{E\}_{n+1}$ in binary form undergo crossover and are mutated to form $\{E\}_{n+3}$; and
8. state $\{E\}_{n+3}$ is then reproduced using the Metropolis criterion.

The genetic MCMC presented in this section is different from the traditional genetic algorithms in the following ways:

- the genetic MCMC does not generate a new population of genes at any given iteration (*i.e.*, generation in the genetic algorithm framework) as is the case in a genetic algorithm, but it generates one sample at each iteration;

- the fitness function uses the Metropolis criterion, while this is not the case in the genetic algorithm; and
- the genetic MCMC has a higher mutation rate than the genetic algorithm.

The random walk in the classical MCMC was replaced by a procedure inspired by Darwin's theory of evolution, which entails crossover, mutation, and reproduction, and it operates in floating-point space.

10.3 Example 1: Simple Beam

The aluminum beam shown in Chapter 2 was used to test the Bayesian finite-element-model-updating methods that were trained using the classical MCMC as well as the genetic programming MCMC. This beam had the following dimensions:

- length = 1.1 m;
- width = 210.2 mm; and
- thickness =10.6 mm.

This beam had holes of diameters 5.8 mm located at the centers of elements 2 to 9 and was therefore difficult to model. Further details of this beam were reported in Marwala (1997). The beam was freely suspended using elastic rubber bands. The beam was excited using an electromagnetic shaker and the response was measured using an accelerometer. The beam was also modeled using a finite-element model as explained in previous chapters, and was divided into 11 elements. The finite-element model used Euler–Bernoulli beam elements. It was excited at various positions and the acceleration was measured at 10 different positions.

A set of 10 frequency-response functions were calculated and a roving accelerometer was used for the testing. Unlike the previous chapters, this chapter used the moduli of elasticity of these elements, densities, and cross-sectional areas as updating parameters.

In applying the Bayesian framework, Equation 10.7 was used and prior information was divided into four parts. Each part had its own coefficient of prior distribution (α_1, α_2, α_3 and α_4). These coefficients are also shown in Equation 10.7 by setting Q equal to 4.

The coefficient α_1 is associated with the density of the beam and was known to be uniform for all elements and was also known to be fairly accurate. The coefficient α_1 was set to 10 to ensure that the density of the beam was not updated significantly.

The coefficient α_2 was associated with the moduli of elasticity of all elements. All elements were known to have uniform modulus of elasticity, which was fairly accurately known. The coefficient α_2 was set to 10 to ensure that the modulus of elasticity was not updated significantly.

The coefficient α_3 was associated with the cross-sectional areas of 9 elements that were known fairly accurately. The coefficient α_3 was set to 10 to ensure that the cross-sectional areas of these elements were not updated significantly. The coefficient α_4 was associated with the cross-sectional areas of 2 elements that were not known accurately because they enclose the area that was drilled to mount the excitation device.

The coefficient α_4 was set to 0.1 to ensure that the cross-sectional areas of these elements were updated significantly. The MCMC method and genetic programming MCMC were implemented by employing the Metropolis acceptance criterion (see Equation 10.11) and 1000 samples were retained to form the posterior probability distribution function indicated by Equation 10.7. The results appear in Table 10.1.

For the genetic part of the simulation, the rate of mutation is 6.6% and the rate of crossover was 70%. It should be noted that the rate of mutation presented here was higher than that of the standard genetic algorithm. The presented Bayesian method via genetic programming has a random component search and, therefore, may be viewed as being equivalent to the random walk that was executed in the standard Bayesian sampling. Indeed, the presented procedure may, in principle, be equivalent to a standard random walk. However, it takes into account the efficient sampling in binary space that was observed in the standard genetic algorithm. It must be noted that the rate of mutation chosen here was lower than the rate of crossover, which is in accordance with many natural systems. When implementing the genetic framework through a genetic algorithm, 16-bit binary numbers were used.

When a modulus of elasticity of 7×10^{10} N m^{-2} was assumed, the error between the first measured natural frequency and that of the initial finite-element model was 1.9%. When the MCMC method was used for finite-element-model updating, this error was reduced to 1.0%, while using the genetic programming MCMC method reduced it to 0.7%.

Table 10.1 Results showing measured frequencies of the beam, the initial frequencies and the frequencies obtained when the finite-element model of a beam was updated using the MCMC and the genetic programming MCMC method

Modes	Measured frequency (Hz)	Initial frequency (Hz)	Frequencies from the MCMC (Hz)	Frequencies from the genetic programming MCMC (Hz)
1	41.5	42.3	41.1	41.2
2	114.5	117.0	116.7	112.4
3	224.5	227.3	222.7	227.7
4	371.6	376.9	374.9	370.5

The error between the second measured natural frequency and that of the initial model was 2.2%. When the MCMC method was used, this error was reduced to 1.9% and in using the genetic programming MCMC it was increased to 1.8%.

The error in the third natural frequencies between the measured data and the initial finite-element model was 1.2%. When the MCMC method was used, this error was reduced to 0.8%, but using the genetic programming MCMC it increased to 1.4%.

The error between the fourth measured natural frequency and that of the initial model was 1.4%. When the MCMC method was used, this error was reduced to 0.9% and using the genetic programming MCMC it was reduced to 0.3%.

Overall, the genetic programming MCMC gave the best results. Both methods improved on average when compared to the average error between the initial finite-element model and the measured data.

The updated models implemented were also validated on the mode shapes they predicted. To make this assessment possible, the modal assurance criterion (MAC) (Allemang and Brown, 1982) was used. The results are shown in Table 10.2. The mean of the diagonal of the MAC vector was used to compare the mode shapes predicted by the updated and initial finite-element models with the measured mode shapes. An average value of 1.0 indicates that the mode shapes were properly correlated. The average MAC calculated between the mode shapes from an initial finite-element model and the measured mode shapes was 0.9986. When the average MAC was calculated between the measured data and data obtained from the updated finite-element models, it was observed that both models give a marginally improved average for the diagonals of the MAC matrices of 0.9989.

Table 10.2 Beam results showing the MAC calculated between measured mode shapes and the initial finite-element model, the MCMC and the genetic programming MCMC

Method	**MAC**
Initial model	0.9986
MCMC	0.9989
Genetic programming MCMC	0.9989

10.4 Example 2: Unsymmetrical H-shaped Structure

The unsymmetrical H-shaped aluminum structure shown in Chapter 2 was also used to validate the proposed method. This structure was also used by Marwala (1997). The structure was excited using an electromagnetic shaker and the response was measured with an accelerometer. The structure was divided into 12 elements, and it was excited with acceleration being measured at 15 positions. The structure was tested freely suspended, and a set of 15 frequency-response functions were calculated. A roving accelerometer was used for the testing. The mass of the accelerometer was found to be negligible compared to the mass of the structure.

As in the previous example, a finite-element model was constructed using the Euler–Bernoulli beam elements. The finite-element model contained 12 elements. As in the previous example, the moduli of elasticity of these elements were used as updating parameters, which were restricted to fall in the interval 6×10^{10} to 8×10^{10} N m^{-2}.

In applying the Bayesian framework, Equation 10.7 was used and prior information was divided into four parts with each part having its own coefficient of prior distribution (α_1, α_2, α_3 and α_4). These coefficients are also shown in Equation 10.7 by setting Q equal to 4.

The coefficient α_1 is associated with the density of the beam and it is known to be uniform for all elements, and is also known to be fairly accurate. The coefficient α_1 was set to 10 to ensure that the density of the beam was not updated significantly. The coefficient α_2 is associated with the moduli of elasticity of all elements.

All elements were known to have a uniform modulus of elasticity, which was known fairly accurately. The coefficient α_2 was set to 10 to ensure that the modulus of elasticity was not updated significantly.

The coefficient α_3 is associated with the cross-sectional areas of 5 elements that were known fairly accurately. The coefficient α_3 was set to 10 to ensure that cross-sectional areas of these elements were not updated significantly.

The coefficient α_4 is associated with the cross-sectional areas of 4 elements that were not accurately known because they enclose the area which was drilled to mount the excitation device. The coefficient α_4 was set to 0.3 to ensure that the cross-sectional areas of these elements were updated significantly.

The MCMC method and the genetic programming MCMC was implemented by employing the Metropolis acceptance criterion (see Equation 10.11). A total of 1000 samples were retained to form a posterior probability distribution function indicated by Equation 10.7. The results appear in Table 10.3.

For the genetic part of the simulation, the rate of the same implementation was conducted as in the previous section. The results obtained when the MCMC method and the genetic programming MCMC were used for finite-element-model updating are shown in Table 10.3. The table shows the measured natural frequencies, initial natural frequencies, and natural frequencies obtained by the MCMC method and the genetic MCMC updated finite-element models.

The error between the first measured natural frequency and that from the initial finite-element model, which was obtained when the modulus of elasticity of 7×10^{10} N m^{-2} was assumed, is 4.3%. When the MCMC method was used for finite-element-model updating, this error was reduced to 3.5%, and the genetic programming MCMC approach reduced this error to 1.9%.

Table 10.3 Results from an unsymmetrical H-shaped structure showing measured frequencies, the initial frequencies and the frequencies obtained when the finite-element model was updated using the MCMC and the genetic programming MCMC

Modes	Measured frequency (Hz)	Initial frequency (Hz)	Frequencies from the MCMC (Hz)	Frequencies from the genetic programming MCMC (Hz)
1	510.9	56.2	52.0	52.9
2	117.3	127.1	118.4	123.3
3	208.4	228.4	202.4	211.9
4	254.0	2610.4	265.5	248.9
5	445.1	452.4	453.6	447.8

The error between the second measured natural frequency and that of the initial model was 8.4%. When the MCMC method was used, the error was reduced to 0.9% and the genetic programming MCMC reduced this error to 5.1%.

The error in the third natural frequencies between the measured data and the initial finite-element model was 9.6%. When the MCMC method was used, this error was reduced to 2.9% and using the genetic programming MCMC, it was reduced to 1.7%.

The error between the fourth measured natural frequency and that of the initial model was 3.7%. When the MCMC method was used, this error was increased to 4.5% and using the genetic programming MCMC, it was reduced to 2.0%.

The error between the fifth measured natural frequency and that of the initial model was 1.6%. When the MCMC method was used, this error was increased to 1.9% and the genetic programming MCMC reduced it to 0.6%. Overall, the genetic programming MCMC gave the best results.

As in the previous example, the updated models implemented were validated on the mode shapes they predicted using the MAC. The results are in Table 10.4. The table shows that the MCMC and the genetic programming MCMC updated finite-element models gave improved averages for the diagonals of the MAC matrices of 0.8408 and 0.8410, respectively.

Table 10.4 H-shaped structure results showing the MAC calculated between measured mode shapes and the initial finite-element model, the MCMC and the genetic programming MCMC

Method	MAC
Initial model	0.8394
MCMC	0.8408
Genetic programming MCMC	0.8410

10.5 Conclusion

In this chapter, the finite-element-model-updating problem was placed in the Bayesian framework. The MCMC method and the genetic programming MCMC were implemented to solve the Bayesian formulated problem and thus update the finite-element model. When these two techniques were tested on a simple beam and an unsymmetrical H-shaped structure, it was observed, on average, that the genetic programming MCMC gave more accurate results than the MCMC method.

10.6 Future Work

This chapter introduced Bayesian formulation for finite-element-model updating. The MCMC and the genetic programming MCMC were used to solve the problem. For further work, other sampling methods such as Gibbs sampling should be used. The conclusions reached in this chapter are highly dependent on the nature of the data used in the analysis. Therefore, further statistical tests should be conducted to ensure that the conclusions reached are not dependent on the data used.

References

Allemang RJ, Brown DL (1982) A Correlation Coefficient for Modal Vector Analysis. In: Proc of the 1st Intl Modal Analysis Conf:01–18

Beck JL, Katafygiotis LS (1998) Updating Models and their Uncertainties. I: Bayesian Statistical Framework. J of Eng Mech 124:455–461

Berahman F, Behnamfar F (2009) Probabilistic Seismic Demand Model and Fragility Estimates for Critical Failure Modes of Un-anchored Steel Storage Tanks in Petroleum Complexes. Probab Eng Mech 24:527–536

Bishop CM (1995) Neural Networks for Pattern Recognition. Oxford University Press, London

Bishop CM (2006) Pattern Recognition and Machine Learning. Springer, New York

Campillo F, Rakotozafy R, Rossi V (2009) Parallel and Interacting Markov Chain Monte Carlo Algorithm. Math and Comput in Simul 79:3424–3433

Ching J, Leu SS (2009) Bayesian Updating of Reliability of Civil Infrastructure Facilities Based on Condition-state Data and Fault-tree Model. Reliability Eng and Syst Saf 94:1962–1974

Dimov IT, Philippe B, Karaivanova A, Weihrauch C (2008) Robustness and Applicability of Markov Chain Monte Carlo Algorithms for Eigenvalue Problems. Appl Math Model 32:1511–1529

Duane S, Kennedy AD, Pendleton BJ, Roweth D (1987) Hybrid Monte Carlo. Phys Letters 195:216–222

Ellis RL (1843) On the Foundations of the Theory of Probabilities. Transactions of the Cambridge Philosoph Soc 8:13–18

Ewins DJ (1995) Modal Testing: Theory and Practice. Research Studies Press, Letchworth

Fonlupt C (2001) Solving the Ocean Color Problem Using a Genetic Programming Approach. Appl Soft Comput 1:63–72

Gallagher K, Charvin K, Nielsen S, Sambridge M, Stephenson J (2009) Markov Chain Monte Carlo (MCMC) Sampling Methods to Determine Optimal Models, Model Resolution and Model Choice for Earth Science Problems. Marine and Petrol Geol 26:525–535

Garcia-Arnau M, Manrique D, Rios J, Rodriguez-Paton A (2006) Initialization Method for Grammar-guided Genetic Programming. Knowledge-Based Syst 20:127–133

Geman S, Geman D (1984) Stochastic Relaxation, Gibbs Distributions and the Bayesian Restoration of Images. IEEE Transact on Patt Analysis and Machine Intell 6:721–741

Gill J (2008) Bayesian Methods: A Social and Behavioural Sciences Approach. Chapman and Hall/CRC, London

Goldberg DE (1989) Genetic Algorithms in Search, Optimization and Machine Learning. Addison-Wesley, Reading

Goodman JB, Lin KK (2009) Coupling Control Variates for Markov Chain Monte Carlo. J of Comput Phys 228:7127–7136

Green PJ (1995) Reversible Jump Markov Chain Monte Carlo Computation and Bayesian Model Determination. Biometrika, 82:711–732

Grieco D, Hogarth RM (2009) Overconfidence in Absolute and Relative Performance: The Regression Hypothesis and Bayesian Updating. J of Econ Psychol 30:756–771

Guyan RJ (1965) Reduction of Stiffness and Mass matrices. Am Inst of Aeronaut and Astronaut 3:380

Hald A (1998) A History of Mathematical Statistics from 1750 to 1930. Wiley, New York

Holland J (1975) Adaptation in Natural and Artificial Systems. University of Michigan Press, Ann Arbor

Hongxing H, Sol H, de Wilde WP (2000) On a Statistical Optimization Method Used in Finite Element Updating. J of Sound and Vib 231:1071–1078

Hum MA, Rue H, Sheehan NA (1999) Block Updating in Constrained Markov Chain Monte Carlo Sampling. Statistics and Probab Letters 41:353–361

Katafygiotis LS, Papadimitrio C, Lam HF (1998) A Probabilistic Approach to Structural Model Updating. Soil Dyn and Earthq Eng 17:495–507

Kendall MG (1949) On the Reconciliation of Theories of Probability. Biometrika 36:101–116

Kosorok MR (2000) Monte Carlo Error Estimation for Multivariate Markov Chains. Statistics and Probab Letters 46:85–93

Kouchakpour P, Zaknich A, Braunl T (2009) Dynamic Population Variation in Genetic Programming. Info Sci 179:1078–1091

Kumar AVA, Balasubramaniam P (2007) Optimal Control for Linear Singular System Using Genetic Programming. Appl Math and Comput 192:78–89

Liang F (2009) Improving SAMC Using Smoothing Methods: Theory and Applications to Bayesian Model Selection Problems. Annals of Statistics 37:2626–2654

Lin J, Ke H, Chien B, Yang W (2007) Designing a Classifier by a Layered Multi-population Genetic Programming Approach. Patt Recog 40:2211–2225

Lones MA, Tyrrell AM (2004) Modelling Biological Evolvability: Implicit Context and Variation Filtering in Enzyme Genetic Programming. Biosyst 76:229–238

Marwala T (1997) A Multiple Criterion Updating Method for Damage Detection on Structures. Master's Thesis, University of Pretoria.

Marwala T (2002) Finite Element Updating Using Wavelet Data and Genetic Algorithm. AIAA J of Aircr 39:709–711

Marwala T (2007a) Computational Intelligence for Modelling Complex Systems. Research India Publications, Delhi

Marwala T (2007b) Bayesian Training of Neural Network Using Genetic Programming. Pattern Recognit Lett 28:1452–1458

Marwala T (2009) Computational Intelligence for Missing Data Imputation, Estimation and Management: Knowledge Optimization Techniques. Information Science Reference Imprint, IGI Global Publications, New York

Marwala T, Mdlazi L, Sibisi S (2007) Finite Element Model Updating Using Bayesian Approach arXiv:0705.2515

McKay RI (2001) Variants of Genetic Programming for Species Distribution Modelling – Fitness Sharing, Partial Functions, Population Evaluation. Ecol Modell 146:231–241

Metropolis N, Rosenbluth AW, Rosenbluth MN, Teller AH, Teller E (1953) Equations of State Calculations by Fast Computing Machines. J of Chem Phys 21:1087–1092

Michalewicz Z (1996) Genetic Algorithms + Data Structures = Evolution Programs. Springer-Verlag, New York

Mottershead JE, Friswell MI (1995) Model Updating in Structural Dynamics: A Survey. J of Sound and Vib 167:347–375

Neal RM (1993) Probabilistic Inference Using Markov Chain Monte Carlo Methods. University of Toronto, Toronto: Tech Report CRG-TR-93-1

Neal RM (2003) Slice Sampling. The Annals of Statistics 31:705–767

O'Neill PD (2002) A Tutorial Introduction to Bayesian Inference for Stochastic Epidemic Models Using Markov Chain Monte Carlo Methods. Math Biosci 180:103–114

Oussaidene M, Chopard B, Pictet OV, Tomassini M (1997) Parallel Genetic Programming and its Application to Trading Model Induction. Parallel Comput 23:1183–1198

Parkins AD, Nandi AK (2004) Genetic Programming Techniques for Hand Written Digit Recognition. Signal Process 84:2345–2365

Reda Taha MM, Lucero J (2004) Damage Identification for Structural Health Monitoring Using Fuzzy Pattern Recognition. Eng Struct 27:1774–1783

Robert CP, Casella G (2004) Monte Carlo Statistical Methods. Springer, New York

Rudolf D (2009) Explicit Error Bounds for Lazy Reversible Markov Chain Monte Carlo. J of Complex 25:11–24
Stigler SM (1990) The History of Statistics: The Measurement of Uncertainty Before 1900. Belknap Press/Harvard University Press, Cambridge
Tan Z (2008) Monte Carlo Integration with Markov Chain. J of Statistical Plan and Infer 138:1967–1980
Tonazzini A, Bedini L (2003) Monte Carlo Markov Chain Techniques for Unsupervised MRF-based Image Denoising. Patt Recog Letters 24:55–64
Tsakonas A (2006) A Comparison of Classification Accuracy of Four Genetic Programming-evolved Intelligent Structures. Info Sci 176:691–724
Vapnik VN (1995) The Nature of Statistical Learning Theory. Springer-Verlag, Berlin
von Mises R (1939) Probability, Statistics, and Truth. Dover Publications Inc, Mineola
Watanabe S, Nakamura A (2009) On-line Adaptation and Bayesian Detection of Environmental Changes Based on a Macroscopic Time Evolution System. In: Proc of the IEEE Intl Conf on Acoust, Speech and Signal Process:4373–4376
Worzel WP, Yu J, Almal AA, Chinnaiyan AM (2009) Applications of Genetic Programming in Cancer Research. The Intl J of Biochem and Cell Biol 41:405–413
Wu JR, Li QS (2004) Finite Element Model Updating for a High-rise Structure Based on Ambient Vibration Measurements. Eng Struct 26:979–990
Yuen K, Katafygiotis LS (2001) Bayesian Time-domain Approach for Modal Updating Using Ambient Data. Probab Eng Mech 16:219–231

Chapter 11

Finite-element-model Updating Applied for Damage Detection

Abstract. This chapter presents a multiple criterion method (MCM) that was tested in damage detection of a simple beam with holes and an irregular H-shaped structure. The MCM was compared with the frequency-response function method (FRFM) and the modal property method (MPM) in terms of their abilities to detect damage in structures. The MCM and FRFM methods were generally found to be able to identify damage better than the MPM.

Keywords: multi-criteria method, modal property method, frequency-response method, damage detection

11.1 Introduction

This chapter uses a multiple criterion method (MCM), a modal property method (MPM), and a frequency-response function method (FRFM) for finite-element model updating, as described in Chapter 8, for damage detection (Marwala and Heyns, 1998). These updating procedures are primarily based on the use of vibration data (Doebling *et al.*, 1996). The use of vibration data for damage detection has been found to be a viable method in the past (Marwala and Hunt, 2000).

Marwala (2001a) introduced the pseudo-modal energy, defined as the integrals of the real and imaginary components of the frequency-response functions over various frequency ranges, for damage identification in structures. Equations that formulate pseudo-modal energies in the modal domain and their respective sensitivities were derived in receptance and inertance form Ewins (1986). When tested on a simulated cantilevered beam, pseudo-modal energies were found to be:

- more resistant to noise in the data than the mode shapes;
- could take into account the out-of-frequency band modes; and
- were better indicators of faults than the modal properties.

Furthermore, they were found to be more sensitive to faults than the natural frequencies and were found to be equally as sensitive to faults as the mode shapes. The pseudo-modal energies were computationally faster to calculate than the modal properties. When tested on a population of 20 steel cylinders, the pseudo-modal energies were, on average, better indicators of faults than the modal properties.

Marwala and Heyns (1998) introduced a multiple-criterion updating method that minimized the error based on modal properties and the frequency-response functions. They applied this method for detecting damage in structures. It was found that the multiple-criterion updating method predicted the presence, the position, and the extent of damage well. When compared with the FRFM and the MPM techniques, the MCM was found to give better results than the other two methods. This was because it could better detect damage to the structure than the modal property method (which failed to detect multiple-damage cases) and gave results that were less noisy, *i.e.*, less updating to undamaged elements was required than for the frequency-response method.

Marwala (2004) introduced a fault-identification technique that used pseudo-modal energies to train probabilistic neural networks. This method was tested on a population of 20 cylindrical shells and its performance was compared to the technique that used modal properties to train probabilistic neural networks. The probabilistic neural networks were trained using pseudo-modal energies provided a better classification of faults than the probabilistic neural networks trained using the conventional modal properties.

Marwala (2003) introduced a fault-identification technique that used pseudo-modal energies to train neural networks. His method was tested on a simulated cantilevered beam and a population of 20 cylindrical shells, and its performance was compared to that of a method that used modal properties to train neural networks. Both the cantilevered beam and cylindrical shells were divided into three substructures, and faults were introduced into these substructures. The cylinder was excited using a modal hammer, and acceleration was measured using an accelerometer. Each fault case was assigned a fault identity, with the presence of a fault represented by a 1, and the absence of a fault was represented by a 0. Following this fault-representation scheme, a fault located in substructure 1 would have an identity of [1 0 0], with the two zeros indicating the absence of faults in substructures 2 and 3. The neural network used was a multi-layer perceptron, trained using the scaled conjugate method. The statistical overlap factor and principal component analysis were used to reduce the size of the input data. For both examples the pseudo-modal-energy-trained neural networks provided a better classification of faults than the networks trained using the conventional modal properties.

Marwala (2001b) applied Bayesian formulated neural networks for probabilistic fault identification in structures. Each of the 20 nominally identical cylindrical shells was arbitrarily divided into three substructures. Holes of 10–15 mm diameter were introduced in each of the substructures and vibration data were measured. Modal properties and the coordinate modal assurance criterion (COMAC), with a natural-frequency vector taken as an additional mode, were used to train the modal-property network and the COMAC network (Friswell and

Mottershead, 1995). Modal energies were calculated by determining the integrals of the real and imaginary components of the frequency-response functions over bandwidths of 12% of the natural frequencies. The modal energies and the coordinate modal energy assurance criterion (COMEAC) were used to train the modal-energy network and the COMEAC network. The average of the modal-property network and the modal-energy network as well as the COMAC network and the COMEAC network formed a modal-energy–modal-property-committee and COMEAC–COMAC committee, respectively. Both committees were observed to give lower mean square errors and standard deviations than their respective individual methods. The modal-energy and COMEAC networks were found to give more accurate fault-identification results than the modal-property network and the COMAC network. For classification (of the presence or absence of faults) the modal-property network was found to give the best results, followed by the COMEAC–COMAC committee. The modal energies and modal properties were observed to give better identification of faults than the COMEAC and the COMAC data. The main advantage of the Bayesian formulation was that it gave the identities of the damage and their respective standard deviations. The use of vibration data for damage detection generally involves the following procedure:

- measurement of vibration data;
- cleaning the data;
- using the vibration data to construct or capacitate the model. Here, the data can be used to build a model, as in neural networks, or to capacitate the existing model, as was the case in finite-element-model updating; and
- given the measured data, diagnosing the damage state of the structure.

11.2 Data Used for Damage Detection

This section describes the data that can be used for damage detection. This chapter used vibration data for fault identification. Vibration data can be presented in three different domains, *i.e.*, the time, frequency and time–frequency domains. There are many damage detection methods that have been proposed and these methods are formulated in these domains.

11.2.1 Time Domain

The time domain is the domain in which data are measured. The time domain is a history that demonstrates how a given signal changes over time. The time-domain data are usually applicable in the following situations:

- when the system in question is time dependent;
- in nonlinear system; and
- broadband-noise propagation can be dealt with without difficulty.

Zimin and Zimmerman (2009) presented a time-domain-based structural-damage-detection method and successfully applied this to simulated and experimental data. Qu and Peng (2007) implanted a time-domain damage-detection technique for the vertical bars of a mast structure. Simulated examples demonstrated that their proposed method could correctly identify the location of the damage.

Majumder and Manohar (2003) presented a time-domain method for damage detection in beam structures with a moving vehicle as an excitation source. A successfully updated finite-element model of an undamaged bridge structure existed. Changes to the updated model that imitated the changes in bridge behavior because of damage were established by using a time-domain method. The effectiveness of the methods presented was shown by taking into consideration the detection of localized and distributed damages.

Lopez III and Zimmerman (2002) applied the time-domain method and the minimum rank perturbation theory for nonlinear damage detection using a time-domain model. The modal minimum rank perturbation theory calculates perturbation matrices by estimating the structural alterations from a linear state to another one because of the damage. This method was successfully tested on a nonlinear, three degrees-of-freedom oscillator and simulated data from a 96-degrees-of-freedom system with simulated noise.

Trickey *et al.* (2002) applied a time-domain technique for damage detection. Their method characterized changes in the geometric properties of the response of a structure and their results, from a finite-element model of a thin plate with damage, showed an increase in the sensitivity to damage when compared to modal-based techniques.

Cattarius and Inman (1997) applied a time-domain method for damage detection in smart structures by constructing two finite-element models to inspect axial and transverse vibrations. The different time responses from different material defects demonstrated the presence of damage in situations of minimal frequency shifts. Other applications of time-domain data for damage detection include the use of data from reflectometry in a packaging system and for electrical time-domain reflectometry in concrete structures, as well as in rock deformation (Lin *et al.*, 1998; Lin *et al.*, 1999). Lew *et al.* (1997) compared the mode shape curvature method, minimum rank update method and transfer function parameter change method to identify the location of damage position.

11.2.2 Frequency Domain

The frequency domain is a representation of how the nature of the signal changes us a function of frequency. The way this process is achieved is by taking a signal in the time domain and then transforming it through the mathematical technique called the *Fourier transform* into the frequency domain (Bochner and Chandrasekharan, 1949; Bracewell, 2000).

Lee and Kim (2007) applied the frequency-domain method for structural damage detection in the frequency domain. A signal index measuring changes in frequency-response functions (FRFs) was presented and successfully applied to detect damage. Fasel *et al.* (2003) successfully applied auto-regressive models with exogenous inputs in the frequency domain for damage detection. Nelwamondo and

Marwala (2006) applied the Mel-frequency ceptral coefficient for damage detection in bearings.

11.2.3 Modal Domain

The other types of data that have been used for damage detection are the modal properties (Marwala, 2001a). Modal properties are usually extracted from the frequency-response functions using a procedure called *modal analysis*. The modal properties are the mode shapes, natural frequencies and the corresponding damping effect. The parameters are sensitive to parameter changes and therefore can be used for damage detection in structures (Marwala and Sibisi, 2005). Shahdin *et al.* (2009) studied the relationship between impact damages and modal properties for carbon fiber. Other applications of modal data for damage detection include:

- pipes (Jian-Hua *et al.*, 2009);
- metallic foams (Dattoma *et al.*, 2008);
- offshore structures (Li *et al.*, 2008); and
- plane-frame structures (Chen, 2008).

11.2.4 Time–Frequency Domain

The time-domain data described in the previous section can also be transformed into the time–frequency domain. The advantage of the time–frequency domain is that both the time and the frequency domains are viewed at the same time. This is applicable for signals whose characteristics are changing as a function of time. There are many different types of time–frequency techniques and these include the wavelet transform, short-time Fourier transform, Gabor distribution and Wigner–Ville distribution (Marwala, 2002). The choice of which one to use is dependent on the application in question. For example, the Wigner distribution function (WDF) has a high clarity because of the auto-correlation function but the consequence of this is a crossterm shortcoming. Consequently, the analysis of a single-term signal by using the WDF is beneficial. However, when the signal has multiple components, the Gabor transform is more preferable (Addison, 1992; Daubechies, 1992).

A *wavelet* is a mathematical function that is applied to split a continuous-time signal into a number of scale components with frequency ranges corresponding to each component. Each component is then analyzed with a resolution corresponding to its scale. The advantage of wavelet techniques (WT) is their ability to characterize functions that have discontinuities and sharp peaks. Furthermore, WTs are able to perfectly deconstruct and reconstruct nonperiodic and nonstationary signals.

Yan *et al.* (2010) applied WTs for damage detection in a process that localized multiple damages and when the damage occurred, while Huang *et al.* (2009) applied a WT, which can use data from a discrete set of nodes and provide spatially continuous variation in the structural response parameters to monitor structural degradation. Gökdağ and Kopmaz (2009) applied a continuous and discrete WT for damage detection in beam-type structures, while Pakrashi *et al.* (2009) applied wavelet, kurtosis and pseudo-fractal techniques individually for damage detection in the presence of noise.

Seker *et al.* (2008) applied a WT for detecting bearing damage in electric motors by using vibration data as features, while Bombale *et al.* (2008) used a WT for detecting damage in composite beams and plates. Other successful applications of a WT for damage detection include its application for:

- a permanent magnet synchronous machine with bearing damage (Rosero *et al.*, 2009);
- composite structures (Ding *et al.*, 2009);
- a thermal-protection system (Jiang *et al.*, 2009); and
- a rectangular plate (Yang *et al.*, 2009).

In this section the domains of the data that may be used for damage detection were described. This chapter used two domains for damage detection: the modal as well as the frequency domain. The next section describes the methods used for damage detection.

11.3 Model Identification Methods

In this chapter the models that use the data described in the previous section for identifying damage are described. These include neural networks, support vector machines, fuzzy logic and rough sets.

11.3.1 Neural Networks

A *neural network* is an information-processing paradigm that is inspired by the way biological nervous systems, like the human brain, process information. It is a computer-based model of the way the brain performs a particular function. It is an exceptionally powerful instrument that has found successful application in mechanical engineering (Vilakazi and Marwala, 2007), civil engineering (Marwala, 2000), biomedical engineering (Mohamed *et al.*, 2005), finance (Patel and Marwala, 2006) and political science (Lagazio and Marwala, 2005).

Park *et al.* (2009) applied neural networks, acceleration signals and modal data for damage detection in beams, while Niu *et al.* (2009) applied time-delay neural networks for damage detection in structures. Other successful applications of neural networks for damage detection include:

- the use of randomized trained neural networks (Haryanto *et al.*, 2009);
- the use of probabilistic neural networks (Jiang and Zhang, 2008);
- the use of radial basis functions for damage detection (Cheng and Qu, 2008);
- damage detection in a medium-density fireboard panel (Long and Rice, 2008);
- damage detection in bridge joints (Mehrjoo *et al.*, 2008); and
- damage detection in truss structures (Sekine and Watanabe, 2008).

11.3.2 Support Vector Machines

Support vector machines are a supervised learning method used mainly for classification and are derived from statistical learning theory. They were first introduced by Vapnik (1998). They have also been extended to include regression, thus resulting in the term *support vector regression* (SVR) (Gunn, 1997).

Pires and Marwala (2004) used support vector machines for option pricing and further extended these machines to use a Bayesian framework, while Gidudu *et al.* (2007) used support vector machines for image classification. Jayadeva and Chandra (2009) used the regularized least-squares fuzzy support vector regression for financial time-series forecasting, while Zhang *et al.* (2006) used support vector regression for online health monitoring of large-scale structures. Other applications of support vector machines for damage detection include:

- helicopter-blade detection system (Pawar and Jung, 2008);
- He and Yan (2007) who applied wavelet support vector machines in structures;
- damage detection in structures under various boundary conditions (Shimada *et al.*, 2006);
- the use of support vector machines and independent component analysis for damage detection (Song *et al.*, 2005); and
- the use of support vector machines and amplitude modulation for damage detection (Mita and Taniguchi, 2004).

11.3.3 Fuzzy Logic

Fuzzy logic is a mathematical technique that is used to convert human knowledge into a computational language. Fuzzy sets are sets that have elements with degrees of membership. In other words, in fuzzy logic an element of a set has a degree of belonging or membership to that particular set. Zadeh (1965) introduced fuzzy sets as an expansion of the classical concept of a set.

In classical set theory, the membership of elements in a set is evaluated in binary terms in that *e.g.*, either it is a member of that set or it is not a member of that particular set. Fuzzy set theory allows for the steady evaluation of the membership of elements in a set and this is implemented with the help of a membership function that is allowed only to fall within the interval [0, 1]. A fuzzy set is therefore a generalized version of a classical set. Conversely, a classical set is a special case of the membership functions of fuzzy sets that only permit the values 0 or 1. Thus far, fuzzy set theory has not generated any results that differ from the results from a probability or classical set theory.

Chandrashekhar and Ganguli (2009) applied a fuzzy logic, with a sliding-window defuzzifier, and mode shape curvatures for damage detection. The modal curvatures that changed because of damage were fuzzified using the Gaussian fuzzy sets and were related to damage identity using the fuzzy logic. Experimental investigations demonstrated that the technique identified damage correctly even in the presence of noise and the absence of some modal data. Other successful applications of fuzzy logic for damage identification include:

- combining oil analysis and vibration using fuzzy logic for gear-damage detection (Dempsey and Afjeh, 2004);
- the use of strain energy mode shapes and fuzzy logic for damage detection (Sazonov *et al.*, 2002); and
- fuzzy logic and continuum damage mechanics to identify damage (Sawyer and Rao, 2000).

11.3.4 Rough Sets

The *rough set* theory was introduced by Pawlak (1991). It is a mathematical tool to deal with vagueness and uncertainty. It is based on a set of rules that are expressed in terms of linguistic variables. Rough sets are of fundamental importance to computational intelligence and in cognitive science. They are very applicable in the tasks of machine learning and decision analysis, especially in the analysis of decisions where there are inconsistencies. Because they are rule-based, rough sets are very transparent. However, they are not as accurate in their predictions, and most certainly are not universal approximators, as other machine-learning tools, such as neural networks, are. It can thus be concluded that in machine learning there is always a trade-off between prediction accuracy and transparency.

Crossingham and Marwala (2007) presented an approach to optimize rough set partition sizes using various optimization techniques. Three optimization techniques were implemented to perform the granularization process, namely the genetic algorithm, hill climbing and simulated annealing. These optimization methods maximize the classification accuracy of the rough sets. The accuracies achieved after optimizing the partitions using genetic algorithm, hill climbing and simulated annealing were 66.89, 65.84 and 65.48%, respectively, compared to an accuracy of equal-width-bin partitioning of 59.86%.

Rough sets theory provides a technique of reasoning from vague and imprecise data (Goh and Law, 2003). The technique is based on the assumption that some information is associated somehow with some information in the universe of the discourse (Komorowski *et al.*, 1999; Kondo, 2006). Objects with the same information are indiscernible in the view of the available information. An elementary set consisting of indiscernible objects forms a basic granule of knowledge. A union of elementary sets is referred to as a *crisp set* otherwise the set is considered a *rough set*.

The use of rough sets for damage detection has not been explored extensively. Some of the limited explorations of rough sets include that by Li *et al.* (2006) who used rough set theory for damage detection in a simple supporting plank. The fiber Bragg-grating strain sensing array was applied and the results showed that the method performed well for damage detection. Hu *et al.* (2003) also used the rough sets method for damage detection.

11.4 Finite-element-updating Approach

In the previous sections some methods for creating a model that can be used for damage identification were explained. In this section one such model, the finite-element model, is documented. In particular, finite-element-model updating, which is the theme for this book, was applied for damage procedure. The process followed was:

1. Create a finite-element model of the structure.
2. Measure the data from the structure.
3. Update the finite-element model in Step 1 to better reflect the data measured in Step 2.
4. After damage is introduced, take measurements.
5. Further update the updated model identified in Step 3 to reflect the measurements identified in Step 4.
6. Use the updated parameters as an indicator of damage *e.g.*, if only element 4 is updated, indicates that the damage is in element 4.

Jaishi and Ren (2006) applied finite-element-model updating for damage detection. An objective function based on the modal flexibility residual was created and its gradient was computed. The updated parameters were then used as indicators of damage. The procedure was first tested numerically on a simply supported beam with added noise and the results were found to be good. Thereafter, the method was implemented on a reinforced-concrete beam and the results were satisfactory. Other examples of the use of finite-element-model updating for damage detection include damage detection:

- on stay cables (Mordini *et al.*, 2008);
- in a steel structure (Wu and Li, 2006);
- in composite structures (Nosenzo *et al.*, 2003);
- in truss structure (Sorohan, 2004); and
- in beams and on a concrete highway (Teughels and De Roeck, 2005).

In this chapter the updating procedures described in Chapter 8 are used for damage identification. The first updating procedure is the frequency-response function method (FRFM) that updates the finite-element model using the frequency-response functions directly. As explained before, this is achieved by minimizing the following equation:

$$e = \left\| \sum_{j=1}^{N} \left\| \left[-\omega_j^2[M] + i\omega_j[C] + [K] \right] \{X(\omega_j) - \{F(\omega_j)\} \right\| \right\| \quad (11.1)$$

Here, N is the number of frequency points; $[M]$ is the mass matrix; $[C]$ is the damping matrix; $[K]$ is the stiffness matrix; $\{X(\omega_j)\}$ is the response vector; and $\{F(\omega_j)\}$ is the force vector, respectively. The parameter ω_j is the jth frequency point.

As an alternative to the measured frequency-response functions, the modal properties may be used for damage identification through the use of a finite-element model. This method is called the modal property method (MPM). The MPM is realized by minimizing the following error function:

$$e=\left\|\sum_{j=1}^{M}\left|\omega_j^2\{\phi\}_j^T[M]\{\phi\}_j-\{\phi\}_j^T[K])\{\phi\}_j\right|\right\| \tag{11.2}$$

Here, $|\ |$ stands for the absolute value; $\|\ \|$ stands for the Euclidean norm; ω_j is the jth natural frequency; $\{\phi\}_j$ is the jth mode shape; and T stands for transpose.

The FRFM and MPM work well except for the fact that they usually give different solutions. To combine the information from both the modal properties and the FRFs, this chapter uses a method combining the frequency and the modal domains. This method is essentially framed as a multi-criterion optimization problem and can be obtained by combining Equations 11.1 and 11.2 with some scaling functions to get the equation:

$$\begin{aligned} e=&\frac{1}{f_1}\left\|\sum_{j=1}^{M}\left|\omega_j^2\{\phi\}_j^T[M]\{\phi\}_j-\{\phi\}_j^T[K])\{\phi\}_j\right|\right\|+\ldots \\ &\frac{1}{f_2}\left\|\sum_{j=1}^{N}\left|\left[-\omega_j^2[M]+i\omega_j[C]+[K]\right]\{X(\omega_j)-\{F(\omega_j)\}\right|\right\| \end{aligned} \tag{11.3}$$

Here,

$$f_1=\left\|\sum_{j=1}^{M}\left|\omega_j^2\{\phi\}_j^T[M]\{\phi\}_j-\{\phi\}_j^T[K])\{\phi\}_j\right|\right\|_0 \tag{11.4}$$

and

$$f_2=\left\|\sum_{j=1}^{N}\left|\left[-\omega_j^2[M]+i\omega_j[C]+[K]\right)\right]\{X(\omega_j)-\{F(\omega_j)\}\right|\right\|_0 \tag{11.5}$$

In Equations 11.4 and 11.5 the subscript 0 indicates the parameters of the initial estimation of the finite-element parameters. The parameters that are used for finite-element updating in this chapter are the moduli of elasticity of each element. To minimize Equations 11.1, 11.2 and 11.3 a hybrid optimization method that combined particle-swarm optimization and the Nelder–Mead simplex method was implemented, as outlined in Chapter 8. As indicated before, the particle-swarm optimization method was used because of its global search capabilities, while the Nelder–Mead search algorithm, described in detail in Chapter 2, was used because of its local search capabilities.

11.5 Example 1: Suspended Beam

The aluminum beam (Specimen 1) used in this section had the following dimensions:

- length: 1.1 m;
- width: 29.2 mm; and
- thickness: 9.6 mm.

The beam had holes of diameter 5.8 mm located at nodes 2 to 9 that were separated by equal 10 cm spacing as shown in Figure 11.1.

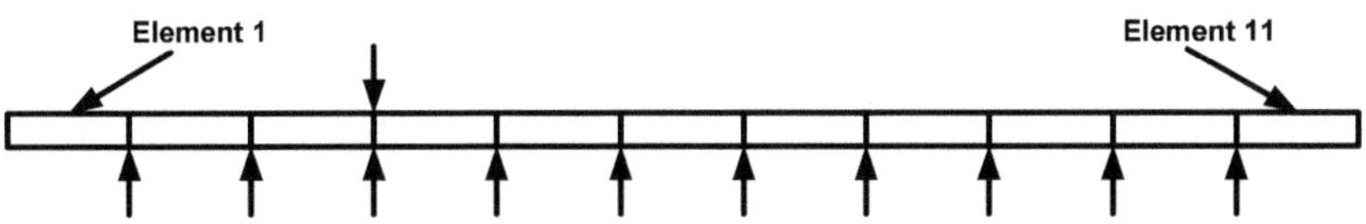

Figure 11.1 The beam with holes that was tested freely suspended

The beam was excited at 10 different positions as shown by the upward arrows, and the FRFs were calculated as in the previous chapters. Furthermore, modal analysis was used to extract the modal properties. A finite-element model was constructed as outlined in previous chapters. When the finite-element model was updated the moduli of elasticity of each of the 11 elements was used as design variables and the results in Table 11.1 were obtained. From Table 11.1 it can be seen that the MPM gave the finite-element model that best predicted the measured data, followed by the MCM and then the FRFM.

Damage case 1 was defined as a case where damage was introduced in element 2. The results obtained are shown in Table 11.2. Table 11.2 demonstrates that the three approaches work well. For all the measured natural frequencies, the MCM gave the best results followed by the MPM. On average, all three methods improved the finite-element models when compared to the average error between the initial finite-element model and the measured data.

For each element, the modulus of elasticity of the newly updated model was subtracted from the modulus of elasticity of the previously updated model and the difference indicated the level of damage in that element.

Damage was introduced at element 2 and the finite-element-model updating method was performed. The modulus of elasticity vector of the finite-element model before damage was compared to the modulus of elasticity of the newly updated finite-element model. The results indicated that updating was performed in elements 2, 4 and 5. However, the most significant updating was performed in element 2. The fact that significant updating was obtained was an indication that damage had occurred and its location was in element 2.

Table 11.1 Table showing natural frequencies for the undamaged case

Mode number	Experimental natural frequencies (Hz)	Initial natural frequency (Hz)	Updated natural frequency (Hz) (FRFM)	Updated natural frequency (Hz) (MPM)	Updated natural frequency (Hz) (MCM)
1	40.9	38.5	41.9	41.3	41.4
2	115.4	106.1	113.0	115.8	113.0
3	224.6	208.8	221.3	221.3	221.3
4	376.1	347.1	377.2	377.2	377.2

Table 11.2 The natural frequencies for damage case 1

Mode number	Experimental natural frequencies (Hz)	Initial natural frequency (Hz)	Updated natural frequency (Hz) (FRFM)	Updated natural frequency (Hz) (MPM)	Updated natural frequency (Hz) (MCM)
1	41.5	37.9	41.9	42.1	41.6
2	114.5	107.7	111.2	116.8	115.4
3	224.5	206.7	226.7	222.3	225.7
4	371.6	351.4	374.9	368.8	371.1

When the MPM was implemented, the difference between the modulus-of-elasticity vector of the updated finite-element model and that before this damage indicated that the MPM had updated elements 2, 4 and 5. It was also observed that element 2 was updated more significantly than the other elements. The results indicated that the MPM showed that element 2 was the location of damage.

When the MCM was applied the difference between the modulus of elasticity vector of the updated finite-element model before and after damage indicated that updating was performed in elements 2, 4, 5, 8 and 9. However, the most significant updating was performed in element 2. The MCM also showed that damage had occurred in element 2.

The FRFM, the MPM and the MCM could all detect the presence of damage and its location (element 2).

For damage case 2, damage was introduced into elements 2 and 3. The results obtained appear in Table 11.3. Table 11.3 shows that all three methods update the first, second and third natural frequencies well. However, on average, the MPM gives the best approximation, followed by the MCM and then the FRFC. The difference between the newly updated model and the updated model before damage indicates that significant updating was performed in elements 2 and 3. The FRFM shows that damage had occurred in elements 2 and 3. When the MPM was used for damage identification, the differences between the newly updated finite-element model parameters and those before indicate that significant damage occurred in elements 2, 3, 9 and 10. This approach could detect the presence of damage in the structure (by virtue of the change in modulus of elasticity). This approach is very weak in locating damage or showing its extent. The approach was unable to display any meaningful detection of damage because there was no way to uniquely determine its location. When the MCM was implemented the results predict that damage has occurred in element 2 and 3. The results demonstrate that the MCM could detect the location and the extent of damage better. This was then followed by the FRFM. The MPM could detect the presence of damage but failed to locate its location.

Table 11.3 Table showing natural frequencies for damage case 2

Mode number	Experimental natural frequencies (Hz)	Initial natural frequency (Hz)	Updated natural frequency (Hz) (FRFM)	Updated natural frequency (Hz) (MPM)	Updated natural frequency (Hz) (MCM)
1	35.4	41.3	38.4	35.9	36.4
2	110.3	113.0	113.3	109.8	112.3
3	209.3	221.9	219.3	210.9	219.3
4	352.7	368.1	359.7	351.8	353.7

The degrees to which each element was updated for two damage cases are shown in Table 11.4.

Table 11.4 Table showing damage cases and their corresponding change in modulus of elasticity. These changes in modulus are multiplied by 10^8 and are in MPA

Element	2	3	4	5	6	7	8	9	10
Case 1 (FRFM)	–50	3	20	–15	2	4	–3	0	10
Case 1 (MPM)	–48	3	20	–10	1	–2	–3	–1	–2
Case 1 (MCM)	–55	–2	15	–15	–2	1	–4	–3	3
Case 2 (FRFM)	–70	–100	–20	–20	–4	–5	–1	–1	2
Case 2 (MPM)	–40	–35	–5	2	2	4	–2	–40	–10
Case 2 (MCM)	–65	–90	2	5	1	–2	1	–1	2

11.6 Example 2: Freely Suspended H-shaped Structure

In this example the irregular H-shaped structure (Specimen 2) shown in Figure 11.2 was used for damage detection. The structure tested in this section was slightly different from the structure used in the previous chapters in that the initial structure was different from the structure tested before and the structure tested in the previous chapter had holes to begin with. This was done as it was felt that it should be more interesting to test the updating procedure on a difficult structure with holes.

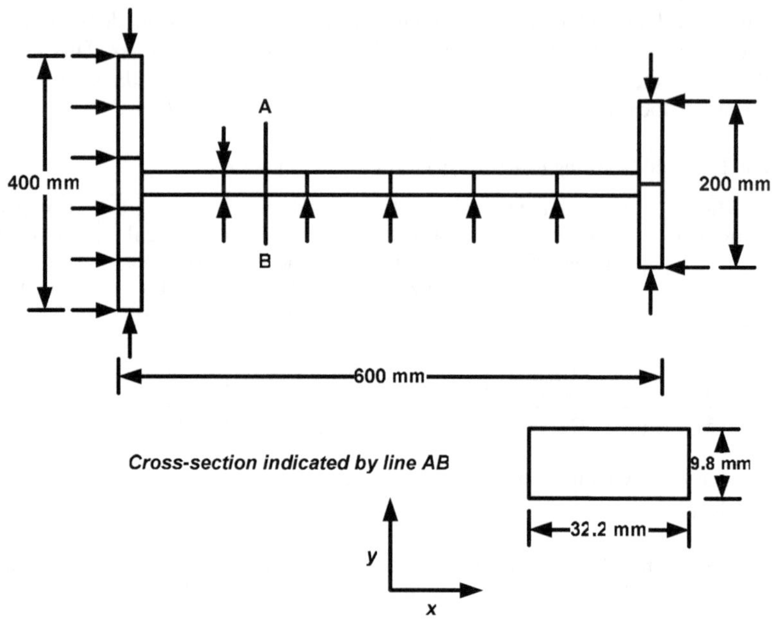

Figure 11.2 Irregular H-shaped structure

The structure displayed in Figure 11.2 was made out of aluminum. The structure was divided into 12 elements. The structure was excited at node 6 and the accelerometer was placed at the 15 locations shown in Figure 11.2 by errors. The structure was tested free suspended and a set of 15 FRFs were obtained and used for updating. The FRFs and the extracted modal properties were used to update the FEM.

Table 11.5 Table showing natural frequencies for undamaged case

Mode number	Experimental natural frequencies (Hz)	Initial natural frequency (Hz)	Updated natural frequency (Hz) (FRFM)	Updated natural frequency (Hz) (MPM)	Updated natural frequency (Hz) (MCM)
1	55.4	51.3	55.3	55.4	55.3
2	125.3	116.0	125.0	125.0	125.3
3	225.2	208.5	226.7	224.7	225.4
4	259.7	240.4	258.7	258.7	259.3
5	446.0	413.0	444.4	444.6	445.6

The initially updated finite-element model was updated using experimental data. The FRFM, MPM and the MCM were used. The results that were obtained are as given in Table 11.5. The table shows a comparison between natural frequencies. Table 11.5 demonstrates that all three approaches were capable of reproducing the measured natural frequencies. The results show that the MPM approximated the measured parameters the best, followed by the MCM and then the FRFM.

For Specimen 1, damage was now introduced into element 3. When finite-element-model updating was performed the results shown in Table 11.6 were obtained.

Table 11.6 demonstrates that the MCM was best able to approximate the natural frequencies, then the MPM, then the FRFM. The results obtained when the FRFM, MPM and MCM were implemented indicated that element 3 was updated more significantly than the other elements. This implies that these methods showed that damage had occurred in element 3.

Table 11.6 The natural frequencies for damage to Specimen 1

Mode number	Experimental natural frequencies (Hz)	Initial natural frequency (Hz)	Updated natural frequency (Hz) (FRFM)	Updated natural frequency (Hz) (MPM)	Updated natural frequency (Hz) (MCM)
1	55.2	55.4	56.0	55.2	54.3
2	123.7	125.3	125.7	123.7	120.1
3	206.8	225.4	208.7	207.9	209.4
4	258.6	259.3	258.5	260.0	257.8
5	443.2	445.6	448.4	442.9	442.9

In Specimen 2, damage was introduced at elements 2 and 3 of the structure. The results obtained when finite-element-model updating was conducted are shown in Table 11.7.

Table 11.7 Table showing natural frequencies in Hz for damage Specimen 2

Mode number	Experimental natural frequencies (Hz)	Initial natural frequency (Hz)	Updated natural frequency (Hz) (FRFM)	Updated natural frequency (Hz) (MPM)	Updated natural frequency (Hz) (MCM)
1	55.2	54.3	55.2	55.5	54.3
2	123.6	120.1	125.7	123.7	120.1
3	205.8	209.4	206.3	207.9	206.9
4	258.5	257.8	260.0	258.5	257.8
5	446.0	442.9	445.6	443.9	442.9

Table 11.7 demonstrates that the FRFM gives best results, then the MCM and then the MPM. In this case damage was introduced in elements 3 and 4. The three approaches were applied as discussed before. The results obtained when the FRFM indicated that elements 3 and 4 were more significantly updated than the other

elements. This approach indicated that damage had occurred in elements 3 and 4. The MPM indicated that elements 2 and 3 were significantly updated. It should be noted, however, that other elements were updated.

The MCM was implemented and the results indicated that damage had occurred in elements 3 and 4. The results demonstrate that the FRFM and the MCM could detect damage to elements 3 and 4.

Although it detected damage on elements 3 and 4, the MPM, did not explicitly detect the location of damage as clearly as the FRFM and MCM. For Case 3 damage was introduced in elements 3, 4 and 5 of the structure. The same procedure as in the previous section was implemented and the results appear in Table 11.8.

Table 11.8 Natural frequencies for damage case 3

Mode number	Experimental natural frequencies (Hz)	Initial natural frequency (Hz)	Updated natural frequency (Hz) (FRFM)	Updated natural frequency (Hz) (MPM)	Updated natural frequency (Hz) (MCM)
1	53.9	54.3	56.3	52.2	52.0
2	117.3	120.1	116.8	119.3	119.1
3	208.4	206.9	213.9	212.1	211.3
4	254.0	257.8	250.3	257.7	256.6
5	445.2	442.9	448.1	443.9	443.9

Table 11.8 shows that all the three methods gave good results, while the results indicating the extent of the updating process in each of the elements appear in Table 11.9.

Table 11.9 Damage case and its corresponding change in modulus of elasticity. These changes in modulus must be multiplied by 10^8 and are in MPA

Element	1	2	3	4	5	6	7	8	9	10	11	12
Case 1 (FRFM)	–1	–21	–98	–5	5	10	8	4	5	–10	–10	–15
Case 1 (MPM)	–25	–7	–160	–18	2	2	1	–1	–3	2	20	–20
Case 1 (MCM)	–10	–55	–2	15	–15	–2	1	–4	–3	3	–10	–7
Case 2 (FRFM)	25	5	–170	–225	4	7	5	4	5	2	5	5
Case 2 (MPM)	–41	–110	–98	–35	–2	3	4	3	4	20	–5	6
Case 2 (MCM)	–5	–20	–135	–83	9	25	2	–2	4	3	1	0
Case 3 (FRFM)	40	2	–187	–213	–150	5	4	4	5	5	2	–4
Case 3 (MPM)	–45	–120	–85	–38	–45	–5	4	–40	–40	–17	–8	–4
Case 3 (MCM)	25	7	–193	–89	–107	8	7	5	4	2	–3	–1

The MCM gave the best updated approximation of the measured natural frequencies, followed by the MPM and then the FRFM. The results indicated that the FRFM updated elements 3, 4 and 5 more significantly than the other elements. When the MPM was employed it seemed to indicate that damage had occurred in elements 2 and 3. The fact that significant updating was performed was an indication that damage was present. When the MCM was implemented, the results indicated that damage had occurred in elements 3, 4 and 5. The results given by the MCM and the FRFM showed that damage had occurred in elements 3, 4 and 5. The MPM could detect the presence of damage but failed to detect its location.

11.7 Conclusion

In this study the MCM, FRFM and MPM techniques were used for damage identification on a beam with holes and on an irregular H-shaped structure. The ability of the MCM to detect damage was compared to the ability of the FRFM and MPM. Generally, the MCM and FRFM methods could detect damage better than the MPM.

11.8 Future Work

The nature of the damage introduced to the structure was a saw-cut. In real structures, the main cause of damage includes fatigue. In fatigue damage, the presence of damage tends to increase the level of damping on the structure. The other issue pertaining to damage involves the location of damage. Damage was generally introduced to one half of the structure only. This was done purposefully to destroy the symmetry of the structure, thereby increasing the probability of the updating method detecting the presence, location and the extent of damage. However, in reality, the presence of damage might not necessarily destroy the symmetry of the structure. Because of this, the proposed updating method needs to be investigated for randomly introduced damage.

References

Addison PS (1992) The Illustrated Wavelet Transform Handbook, Institute of Physics

Bochner S, Chandrasekharan K (1949) Fourier Transforms. Princeton University Press, New Jersey

Bombale BS, Singha MK, Kapuria S (2008) Detection of Delamination Damage in Composite Beams and Plates Using Wavelet Analysis. Struct Eng and Mech 30:699–712

Bracewell RN (2000) The Fourier Transform and Its Applications. McGraw-Hill, Boston

Cattarius J, Inman DJ (1997) Time Domain Analysis for Damage Detection in Smart Structures. Mech Syst and Signal Process 11:409–423

Chandrashekhar M, Ganguli R (2009) Structural Damage Detection Using Modal Curvature and Fuzzy Logic. Struct Health Monit 8:267–282

Chen HP (2008) Application of Regularization Methods to Damage Detection in Large Scale Plane Frame Structures Using Incomplete Noisy Modal Data. Eng Struct 30:3219–3227

Cheng L, Qu W (2008) Structural Damage Detection Based on EMD Method and RBF Neural Network. Earthq Eng and Eng Vib 28:131–138

Crossingham B, Marwala T (2007) Using Optimisation Techniques to Granulise Rough Set Partitions. In Comput Models for Life Sci, Am Inst of Phys 952:248–257

Dattoma V, Giannoccaro NI, Messina A, Nobile R (2008) Fatigue Damage Evaluation of Metallic Foam Specimens through Modal Data. In: Proc of the 11th Intl Congr and Exhibit on Exper and Appl Mech:527–535

Daubechies I (1992) Ten Lectures on Wavelets. Soc for Indust and Appl Maths

Dempsey PJ, Afjeh AA (2004) Integrating Oil Debris and Vibration Gear Damage Detection Technologies Using Fuzzy Logic. J of the Am Helicopter Soc 49:109–116

Ding Y, Yang Z, Wang L, Dang X (2009) Damage Detection of Composite Structure Using Jointly Response Energy and Continuous Wavelet Transform (CWT). J of Northwestern Polytech Univ 27:316–320

Doebling SW, Farrar CR, Prime MB, Shevitz DW (1996) Damage Identification and Health Monitoring of Structural and Mechanical Systems From Changes in Their Vibration Characteristics: A Literature Review. Los Alamos National Laboratory Report LA-13070-MS

Ewins DJ (1986) Modal Testing: Theory and Practice. Research Studies Press, Letchworth

Fasel TR, Sohn H, Park G, Farrar CR (2003) Application of Frequency Domain ARX Models and Extreme Value Statistics to Impedance-based Damage Detection. Am Soc of Mech Eng, Aerosp Div (Publication) AD, 68:289–297

Friswell MI, Mottershead JE (1995) Finite Element Model Updating in Structural Dynamics, Kluwer Academic Publishers, Norwell

Gidudu A, Hulley G, Marwala T (2007) Image Classification Using SVMs: One-against-one vs One-against-all. In: Proc of the 28th Asian Conf on Remote Sens:CD-ROM

Goh C, Law R (2003) Incorporating the Rough Sets Theory into Travel Demand Analysis. Tour Manage 24:511–517

Gökdağ H, Kopmaz O (2009) A New Damage Detection Approach for Beam-type Structures Based on the Combination of Continuous and Discrete Wavelet Transforms. J of Sound and Vib 324:1158–1180

Gunn SR (1997) Support Vector Machines for Classification and Regression. Technical Report, Image Speech and Intell Syst Res Group, University of Southampton

Haryanto I, Setiawan JD, Budiyono A (2009) Structural Damage Detection Using Randomized Trained Neural Networks. Studies in Comput Intell 192:245–255

He H, Yan W (2007) Application of Wavelet Support Vector Machine to Structural Damage Detection. J of Vib, Measur and Diag 27:53–57

Hu SS, Xu DY, Liu Y (2003) Damage Fault Detection of Fighter Based on Rough Set Model. Contr Theor and Appl 20:943

Huang Y, Meyer D, Nemat-Nasser S (2009) Damage Detection with Spatially Distributed 2D Continuous Wavelet Transform. Mech of Mater 41:1096–1107

Jaishi B, Ren WX (2006) Damage Detection by Finite Element Model Updating Using Modal Flexibility Residual. J of Sound and Vib 290:369–387

Jayadeva RK, Chandra S (2009) Regularized Least Squares Fuzzy Support Vector Regression for Financial Time Series Forecasting. Expert Syst with Appl 36:132–138

Jian-Hua D, Zhiwei O, Yangdong Z (2009) Damage Identification Method with Structural Frequency Data from Modal Test. Komunikacie 11:50–54

Jiang SF, Zhang S (2008) Data-fusion Damage Detection Approach Based on Probabilistic Neural Network Classifier. Chinese J of Comput Mech 25:700–705
Jiang X, Mahadevan S, Guratzsch R (2009) Bayesian Wavelet Methodology for Damage Detection of Thermal Protection System Panels. AIAA J 47:942–952
Komorowski J, Pawlak Z, Polkowski L, Skowron A (1999) A Rough Set Perspective on Data and Knowledge. The Handbook of Data Mining and Knowledge Discovery, Oxford University Press, London
Kondo M (2006) On the Structure of Generalized Rough Sets. Info Sci 176:589–600
Lagazio M, Marwala T (2005) Assessing Different Bayesian Neural Network Models for Militarized Interstate Dispute. Social Sci Comput Rev:1–12
Lee J, Kim S (2007) Structural Damage Detection in the Frequency Domain Using Neural Networks. J of Intell Mater Syst and Struct 18:785–792
Lew J, Sathananthan S, Gu Y (1997) Comparison of Damage Detection Algorithms Using Time Domain Data. In: Proc of the Intl Modal Anal Conf – IMAC, 1:645–651
Li H, Wang J, James Hu SL (2008) Using Incomplete Modal Data for Damage Detection in Offshore Jacket Structures. Ocean Eng 35:1793–1799
Li Y, Jiang DS, Luo P, Wang LX (2006) Structural Damage Detection Using Rough Set Theory. J of Wuhan Univ of Technol 28:90–93
Lin MW, Abatan AO, Zhang W (1999) Crack Damage Detection of Concrete Structures Using Distributed Electrical Time Domain Reflectometry (ETDR) Sensors. In: Proc of SPIE - The Intl Soc for Optic Eng 3671:297–304
Lin MW, Abatan AO, Zhang W (1998) Crack Damage Detection of Structures Using Distributed Electrical Time Domain Reflectometry (ETDR) Sensors. In: Proc of SPIE - The Intl Soc for Optic Eng 3325:173–180
Long W, Rice RW (2008) Detection of Structural Damage in Medium Density Fiberboard Panels Using Neural Network Method. J of Compos Mater 42:1133–1145
Lopez III FP, Zimmerman DC (2002) Nonlinear Damage Detection Using a Time-domain Minimum Rank Perturbation Theory. In: Proc of SPIE - The Intl Soc for Optic Eng 4702:179–190
Marwala T (2000) On Damage Identification Using a Committee of Neural Networks. J of Eng Mech 126:43–50
Marwala T (2001a) Fault Identification Using Pseudomodal Energies and Modal Properties. AIAA J 39:1608–1617
Marwala T (2001b) Probabilistic Fault Identification Using Vibration Data and Neural Networks. Mech Syst and Signal Process 15:1109–1128
Marwala T (2002) Finite Element Updating Using Wavelet Data and Genetic Algorithm. J of Aircr 39:709–711
Marwala T (2003) Fault Classification Using Pseudomodal Energies and Neural Networks. AIAA J 41:82–89
Marwala T (2004) Fault Classification Using Pseudomodal Energies and Probabilistic Neural Networks. J of Eng Mech 130:1346–1355
Marwala T, Heyns PS (1998) Multiple Criterion Method for Determining Structural Damage. AIAA J 36:1494–501
Marwala T, Hunt HEM (2000) Is Damage Identification Using Vibration Data in a Population of Cylinders Feasible? J of Sound and Vib 237:727–732
Marwala T, Sibisi S (2005) Finite Element Model Updating Using Bayesian Framework and Modal Properties. J of Aircr 42:275–278
Majumder L, Manohar CS (2003) A Time-domain Approach for Damage Detection in Beam Structures Using Vibration Data with a Moving Oscillator as an Excitation Source. J of Sound and Vib 268:699–716
Mehrjoo M, Khaji N, Moharrami H, Bahreininejad A (2008) Damage Detection of Truss Bridge Joints Using Artificial Neural Networks. Expert Syst with Appl 35:1122–1131

Mita A, Taniguchi R (2004) Active Damage Detection Method Using Support Vector Machine and Amplitude Modulation. In: Proc of SPIE–The Intl Soc for Optic Eng, 5391:21–29

Mohamed N, Rubin DM, Marwala T (2005) Detection of Epileptiform Activity in Human EEG Signals Using Bayesian Neural Networks. In: Proc of the IEEE 3rd Intl Conf on Comput Cybern:231–237

Mordini A, Savov K, Wenzel H (2008) Damage Detection on Stay Cables Using an Open Source-based Framework for Finite Element Model Updating. Struct Health Monit 7:91–102

Nelwamondo FV, Marwala T (2006) Fault Detection Using Gaussian Mixture Models, Mel-frequency Ceptral Coefficient and Kurtosis. In: Proc of the IEEE Intl Conf on Syst, Man and Cybern:290–295

Niu L (2009) A Time-delay Neural Networks Architecture for Structural Damage Detection. In: Proc of the Intl Conf on Adv Comput Contr:571–575

Nosenzo G, Whelan MP, Dalton T (2003) Damage Detection in Composite Structures Based on Optical Fibre Strain Sensing and Finite Element Model Updating. Key Eng Mater 245-346:509–516

Pakrashi V, O'Connor, A, Basut B (2009) A Comparative Analysis of Structural Damage Detection Techniques by Wavelet, Kurtosis and Pseudofractal Methods. Struct Eng and Mech 32:489–500

Park JH, Kim JT, Hong DS, Ho DD, Yi JH (2009) Sequential Damage Detection Approaches for Beams Using Time-modal Features and Artificial Neural Networks. J of Sound and Vib 323:451–474

Patel P, Marwala T (2006) Neural Networks, Fuzzy Inference Systems and Adaptive-Neuro Fuzzy Inference Systems for Financial Decision Making. Lect Notes in Comput Sci 4234:430–439

Pawar PM, Jung SN (2008) Support Vector Machine Based Online Composite Helicopter Rotor Blade Damage Detection System. J of Intell Mater Syst and Struct 19:1217–1228

Pawlak Z (1991) Rough Sets: Theoretical Aspects of Reasoning About Data. Kluwer Academic Publishers, Dordrecht

Pires MM, Marwala T (2004) Option Pricing Using Neural Networks and Support Vector Machines. In: Proc of the IEEE Intl Conf on Syst, Man and Cybern:1279–1285

Qu W, Peng Q (2007) Damage Detection Method for Vertical Bars of Mast Structure in Time Domain. Earthq Eng and Eng Vib 27:110–116

Rosero J, Romeral L, Rosero E, Urresty J (2009) Fault Detection in Dynamic Conditions by Means of Discrete Wavelet Decomposition for PMSM Running Under Bearing Damage. In: Proc of the IEEE Applied Power Electronics Conf and Expo:951–956

Sawyer JP, Rao SS (2000) Structural Damage Detection and Identification Using Fuzzy Logic. AIAA J 38:2328–2335

Sazonov ES, Klinkhachorn P, Gangarao HVS, Halabe UB (2002) Fuzzy Logic Expert System for Automated Damage Detection from Changes in Strain Energy Mode Shapes. Nondestr Test and Eval 18:1–20

Seker S, Güllülü AE (2008) Transfer Function Approach Based Upon Wavelet Transform for Bearing Damage Detection in Electric Motors. In: Proc of the IEEE Intl Symp on Indust Electr:749–752

Sekine H, Watanabe M (2008) Damage Detection of Truss Structures Using a Neural Network with Relearning Process. Trans of the Japan Soc of Mech Eng, Part A, 74:506–512

Shahdin A, Mezeix L, Bouvet C, Morlier J, Gourinat Y (2009) Monitoring the Effects of Impact Damages on Modal Parameters in Carbon Fiber Entangled Sandwich Beams. Eng Struct 31:2833–2841

Shimada M, Mita A, Feng MQ (2006) Damage Detection of Structures Using Support Vector Machines Under Various Boundary Conditions. In: Proc of SPIE – The Intl Soc for Optic Eng, 6174 II:Art No 61742K

Song H, Zhong L, Han B (2005) Structural Damage Detection by Integrating Independent Component Analysis and Support Vector Machine. Lect Notes in Comput Sci:670–677

Sorohan Şt (2004) Finite Element Model Updating and Damage Detection on a Truss Structure. UPB Sci Bullet, Ser D: Mech Eng 66:47–62

Teughels A, De Roeck G (2005) Damage Detection and Parameter Identification by Finite Element Model Updating. Arch of Comput Methods in Eng 12:123–164

Trickey S, Todd M, Seaver M, Nichols J (2002) Geometric Time Domain Methods of Vibration Based Damage Detection. In: Proc of SPIE – The Intl Soc for Optic Eng 4702:113–121

Vapnik V (1998) Statistical Learning Theory. Wiley-Interscience, New York

Vilakazi B, Marwala T (2007) Incremental Learning and Its Application to Bushing Condition Monitoring. Lect Notes in Comput Sci 4491:1241–1250

Wu JR, Li QS (2006) Structural Parameter Identification and Damage Detection for a Steel Structure Using a Two-stage Finite Element Model Updating Method. J of Construct Steel Res 62:231–239

Yan G, Duan Z, Ou J, De Stefano A (2010) Structural Damage Detection Using Residual Forces Based on Wavelet Transform. Mech Syst and Signal Process 24:224–239

Yang JM, Hwang CN, Ke CH, Tseng WM (2009) Damage Detection of a Rectangular Plate with Stiffeners by Two-dimensional Discrete Wavelet Transforms. J of Taiwan Soc of Naval Archit and Marin Eng 28:29–40Zadeh LA (1965) Fuzzy Sets. Info and Contr 8:338–353

Zadeh LA (1965) Fuzzy sets. Information and Control 8:338–353

Zhang J, Sato T, Iai S (2006) Support Vector Regression for On-line Health Monitoring of Large-scale Structures. Structural Safety 28:392–406

Zimin VD, Zimmerman DC (2009) Structural Damage Detection Using Time Domain Periodogram Analysis. Struct Health Monit 8:125–135

Chapter 12

Conclusions and Emerging State-of-the-Art

12.1 Introduction

This book dealt with the use of computational intelligence methods for finite-element-model updating. Finite-element-model updating is a process through which finite-element models are tuned to better reflect the measured data. This is based on the rational assumption that the measured data are more reliable than the finite-element-model's predicted data.

The computational intelligence methods applied in this book are divided into three classes:

- optimization methods;
- machine-learning methods; and
- Monte-Carlo-based methods.

The optimization methods that were used in this book included the Broyden–Fletcher–Goldfarb–Shanno (BFGS), Nelder–Mead (NM) simplex method, genetic algorithm, particle-swarm optimization, simulated annealing, response-surface method and hybrid methods. The reason why optimization methods were used is that the finite-element-model updating problem is essentially an optimization problem where the objective is to minimize the distance between the measured data and the finite-element-model predicted data.

The optimization techniques were mainly global optimum methods with the exception of the BFGS and NM methods, which were local methods. For global optimization the objective was to find the shortest distance between measurements and predictions in the presence of measurement noise.

The machine-learning methods used in this book were multi-layer perceptron neural networks. In the case of the response-surface method, they were used to estimate the relationship between the updating parameters and the differences between the measurements and the prediction, which was minimized. In the second instance, they were used to estimate the relationship between the updating parameters and the natural frequencies.

The other methods used in this book were the Monte Carlo methods. In this book the Monte Carlo methods were used to train a sample probability distribution that was obtained by formulating the problems using the Bayesian framework. In the first instance the hybrid Monte Carlo methods were used to train the Bayesian neural networks, while the traditional Markov chain Monte Carlo (MCMC) and the genetic MCMC methods were used to update the finite-element-model formulated using the Bayesian approach directly.

12.2 Overview of the Previous Chapters

In Chapter 2 the Nelder–Mead simplex method and the Broyden–Fletcher–Goldfarb–Shanno (BFGS) method were applied for the finite-element-model-updating process. These methods were tested on a simple beam and an unsymmetrical H-shaped structure to update their finite-element models to better reflect the measured data. The measured data that were used in this chapter were the natural frequencies and the mode shapes. It was observed that the Nelder–Mead simplex method yielded a finite-element model that was better at predicting measured data than the BFGS. This was mainly because the BFGS method entailed estimating gradients, which was subject to numerical errors.

Chapter 3 compared the Nelder–Mead simplex method, which was found to be better than the BFGS method in Chapter 2, to a genetic algorithm for finite-element model updating. The genetic algorithm (GA) is a global optimum method inspired by the evolutionary process in different species. When the GA was tested, it was found to give a more accurate updated finite-element model than the Nelder–Mead (NM) simplex method did.

In Chapter 4 the particle-swarm-optimization (PSO) method was applied for finite-element-model updating. When this method was tested and compared to a GA it performed better than the GA.

Chapter 5 implemented simulated annealing (SA) for a finite-element-model updating and it was compared to the PSO. It was observed that, on average, the PSO gave a more accurately updated finite-element results than the SA did.

Chapter 6 presented the response-surface method (RSM) for finite-element-model updating. The response-surface method was implemented by approximating the finite-element surface-response equation using a multi-layer perceptron and the updated parameters of the finite-element model were calculated using the GA by optimizing the surface-response equation. The method was compared to existing methods that use simulated annealing and a genetic algorithm separately with a full finite-element model for model updating. The presented method was tested and was found to give similar results to the simulated annealing and the genetic algorithm.

Chapter 7 introduced the hybrid of particle-swarm optimization and the Nelder–Mead simplex optimization method for finite-element-model updating. It was observed that the hybrid method gave more accurate results than when the particle-swarm-optimization or the Nelder–Mead simplex methods were used in isolation.

Chapter 8 introduced a multiple criterion method (MCM) for finite-element-updating model updating. The MCM minimized the Euclidean norm of the error matrix that combined the modal property data and the frequency-response function data. The results given by the MCM were compared to the results from the modal property method (MPM) and frequency-response function method (FRFM). It was observed that the MCM gave the best results followed by the MPM.

Chapter 9 implemented Bayesian neural networks for finite-element-model updating and compared this to the response-surface method. It was found that the Bayesian neural-network method performed better than the response-surface technique.

Chapter 10 implemented the Bayesian approach for finite-element-model updating. The Bayesian approach was implemented using a genetic Markov chain Monte Carlo method inspired by genetic-programming techniques. It was observed that the Bayesian approach gave results that were more accurate on average than the response-surface method.

In Chapter 11 a multiple criterion method (MCM) was presented and tested. The MCM was compared to the frequency-response function method (FRFM) and the modal property method (MPM). The ability of the MCM to detect damage was compared to the ability of the FRFM and MPM. The MCM and FRFM methods were generally found to be better at identifying damage than the MPM was.

12.3 Outstanding Issues

12.3.1 Model Selection

In this book *model selection* was the process of choosing an updated finite-element model from a group of plausible updated finite-element models. Fundamentally, this is essentially about the realization that finding the theory from a series of experimental observations is frequently connected straightforwardly to a mathematical model for estimating those experimental observations.

The aim of model selection was to select the correct finite-element model from the infinite number of plausible finite-element models that might have generated the data. The mathematical technique usually pursued determines a group of plausible finite-element models. Burnham and Anderson (2002) highlight the significance of choosing a group of models based on a logical scientific basis for modeling.

When the group of plausible finite-element models has been chosen, the mathematical approach permitted the most plausible of these models to be decided. The ability to fit the data is usually established by using a likelihood ratio technique. A good model-selection process strikes a balance between the ability of the model to reproduce data with simplicity. The complexity of the model is usually estimated from the number of parameters to be updated in the model.

Complicated models can fit the data, but the additional parameters may not stand for anything functional.

The method of selecting a model is essentially the formation of an approximation of the probability of the model predicting the observed data, while the bias and variance are both significant measures of the quality of this model.

Mthembu *et al.* (2009a) proposed the use of the Bayesian evidence statistics to evaluate the probability of each updating model to assess the condition for choosing the updating parameters in the updating process. The model evidences were compared using the Bayes factor and the Jeffrey scale was used to establish the models differences. The Bayesian evidence was estimated by integrating the likelihood of the data for a given model and the corresponding parameters over the *a priori* model parameter space using the new nested sampling algorithm. The nested algorithm sampled the likelihood distribution, while giving posterior samples of the updating model parameters.

Mthembu *et al.* (2009b) presented the application of particle-swarm optimization (PSO) for updated finite-element-model selection. Each candidate model was characterized as a particle that shows both individualistic and group behavior. Each particle navigated in the model space in search of the best updated finite-element model. An optimal model was described as the model that possessed the lowest number of updated parameters and had the smallest parameter variable variation from the mean material properties. Some of the model selections methods that still need to be explored in the context of the finite-element updating problem should include:

- Akaike information criterion;
- Bayesian information criterion; and
- crossvalidation.

12.3.2 Objective Function

In this book the finite-element-model updating problem was mainly framed as an optimization problem. In optimization, one of the most important steps is to define the objective function that essentially entails the distance between the measured data and the finite-element-predicted data given the updating parameters. There are many different ways in which this objective can be framed. One way is to frame the objective function as a Euclidean norm between the measured data vector and the predicted data vector inclusive of the mode shapes and the natural frequencies. Alternatively, this can be framed as a multi-objective optimization where the two objectives are the Euclidean distance between the natural frequencies vector between measurements as well as predictions and the Euclidean distance between the mode shape vectors between measurements and predictions. Alternatively, other distance measures that can be used instead of the Euclidean norms are the (Mahalanobis, 1936; Cantrell, 2000):

- Mahalabonis distance;
- Manhattan distance; and
- Chebyshev distance.

Within the finite-element-model-updating problem it is not very clear which distance is preferable for a successful process.

12.3.3 Data Used for Finite-element-model Updating

In this book the data that were used for finite-element-model updating were the mode shapes and the natural frequencies, and later on the frequency-response function directly. Other data types can be used for finite-element-model updating. These include wavelet data, modal strain energy and pseudo-modal energies. It is not yet clear which data type is ideal for a given finite-element-model updating procedure. The logical step is to create a unified approach that can update a finite-element model to predict measured data, irrespective of the nature of the presentation of the data. The multi-criteria updating procedure presented in Chapter 8 was a significant step forward but it considered only the modal-domain and the frequency–domain data. Other domains such as the time domain and the time–frequency domain should also be considered to improve the updating procedure.

12.3.4 Local Versus Global Optimally Updated Model

An issue with optimization methods is whether to identify the global optimum solution or not. A *global optimization* in this book implied that the error between the finite-element model and the measured data was at the minimum point. This book mainly used global optimization methods for finite-element-model updating. If the data was completely noise-free and all the possible modes can be measured, then the globally optimized updated finite-element model will be desirable. However, the presence of noise in the data and the fact that that not all possible modes can be measured, imply that single-mindedly pursuing a global optimum solution can often be undesirable. This is because a global optimum case can possibly include modeling the error in the data into the updated finite-element model. The good news is that the physical constraints of the finite-element model imply that unlike regression in statistical models, in the finite-element-model updating process the global optimum solution is desirable. However, work on the significance of the global optimum solution for the finite-element-model updating problem still needs to be explored further.

12.3.5 Online Finite-element-model Updating

Very often, the finite-element-model updating is used for diagnosis in an assembly-line situation and therefore it has to be performed online. In this setting, the finite-element-model updating needs to be performed over a population of nominally identical structures. The theory behind finite-element-model updating online is similar to that offline. However, the difference is that if it is performed online then the issue of computational speed becomes important. For future work, techniques for the online finite-element-model updating process should be implemented using methods such as (Borodin and El-Yaniv, 1998):

- re-enforcement learning;
- dynamic programming;
- temporal difference learning; and
- Q learning.

The success of online finite-element-model updating is highly dependent on the computational efficiency of the updating method. In this chapter an efficient response-surface method based on the multi-layer perceptron was used. In the future, other neural-network methods such as the radial basis functions should be used.

12.3.6 The Issue of Damping

In this book, the finite-element procedures outlined neglected the issue of damping. The reason for this is while damping is an essential concept of structural dynamics it is hard to model. Consequently, the majority of finite-element-model-updating methods proposed thus far simply ignored damping or assumed it to be light and therefore treated it in a simplistic form. For future research, complicated damping models should also be identified as part of the finite-element-model updating process.

12.3.7 Dealing with Nonlinearity

The issue of nonlinearity should be dealt with in finite-element-model updating. In this book, the structures that were analyzed were assumed to be mathematically represented using second order linear differential equations. However, this is a simplistic view of the world, particularly when dealing with problems that are highly nonlinear. In future studies, methods that treat the finite-element-modeling problem as a nonlinear problem should be explored.

12.3.8 Nonuniqueness

The fact that there is more than one updated finite-element model that can reproduce the measured data sufficiently accurately is what is termed *nonuniqueness.* The class of inverse problems such as finite-element-model updating where the measured data are used to identify the correct finite-element model is fundamentally nonunique. In this chapter, the introduction of the Bayesian approach of Chapter 10 was partially aimed at dealing with this issue. Regularization methods have been used to deal with this particular problem (Krause, 1987). The introduction of the multi-criteria method for updating finite-element models introduced in Chapter 8 was also intended to deal with this particular deficiency. In the future, to deal with the problem of uniqueness of the updated finite element the techniques outlined below should be introduced.

Using data presented in multiple domains to rule out implausible models. The general principle here would be the fact that if the updated finite-element model fails to reproduce data in one domain then that updated model cannot be viewed as being universally correct.

The use of the Bayesian approach with the regularization technique framed as a prior probability distribution. In this way, instead of identifying a correct updated finite-element model, then the updating finite-element model is defined probabilistically over the updating parameter space.

12.3.9 Parameter Selection

The issue of which parameters to choose for the updating process still remains arbitrary, despite major advances made in research. The model-selection method is but one of the techniques that can be used for choosing the updating parameters. To date, this problem largely depends on the engineering judgment of the user. With advances in computational intelligence, particularly in areas of teaching computers to behave like human beings, now is perhaps the ideal time to explore the issues of automatic selection of updating parameters.

References

Burnham KP, Anderson DR (2002) Model Selection and Multimodel Inference: A Practical-Theoretic Approach, 2nd edn. Springer-Verlag, Berlin

Borodin A, El-Yaniv R (1998) Online Computation and Competitive Analysis. Cambridge University Press, London

Cantrell CD (2000) Modern Mathematical Methods for Physicists and Engineers. Cambridge University Press

Krause U (1987) Hierarchial structures in multicriteria decision making, in J Jahn, W Krabs (eds) Recent Advances and Historical Development of Vector optimization Springer-Verlag, Berlin:183–193

Mahalanobis PC (1936) On the Generalised Distance in Statistics. In: Proc. Nat. Inst. of Sciences of India 2:49–55

Mthembu L, Marwala T, Friswell MI, Adhikari S (2009a) Bayesian Evidence for Finite Element Model Updating. In: Proc. of the IMAC XXVII, Orlando, Florida

Mthembu L, Marwala T, Friswell MI, Adhikari S (2009b) Finite Element Model Selection Using Particle Swarm Optimization. arXiv:0910.2217.

Appendix A

Finite-element Modeling

A.1 Introduction

In this book finite-element modeling was viewed as the mathematical and numerical processes through which a physical structure was translated into a mathematical model. From that mathematical model a numerical procedure was used to estimate dynamic characteristics such as mode shapes and natural frequencies (Friswell and Mottershead, 1995). The process followed in this book for finite-element modeling entailed (Zienkiewicz, 1986):

- discretization and shape functions;
- estimation of mass and stiffness matrices;
- eigenvalues and eigenvectors estimation; and
- frequency-response function.

A.2 Discretization and Shape Functions

When using finite-element modeling, the shape functions were used to represent the coordinates and the displacement of a position in the finite-element model. For the situation where the coordinates of a position are represented by (x, y, z) and the displacement by (u, v, w), the following expressions can be written (Friswell and Mottershead, 1995):

$$x = \sum_{i=1}^{l} N_j x_j \tag{A.1}$$

$$y = \sum_{i=1}^{l} N_j y_j \tag{A.2}$$

$$z = \sum_{i=1}^{l} N_j z_j \tag{A.3}$$

$$u = \sum_{i=1}^{l} N_j u_j \tag{A.4}$$

$$v = \sum_{i=1}^{l} N_j v_j \tag{A.5}$$

$$w = \sum_{i=1}^{l} N_j w_j \tag{A.6}$$

Here, x, y and z with subscripts j are the coordinates of the jth node and u, v and w is the displacement of this node. The summations in these equations are taken over r nodes, and N_j is the jth node's shape function. The shape functions can be related to the local coordinates (ξ_1, ξ_2, ξ_3) and each surface of the cube will take a constant value of ±1. For the process of shape function formulation to work the following relationship can be defined for the coordinate k:

$$N_k(\xi_{ik}, \xi_{2k}, \xi_{3k}) = 1 \tag{A.7}$$

$$N_j(\xi_{ik}, \xi_{2k}, \xi_{3k}) = 0,\ j \neq k \tag{A.8}$$

There are many elements that can be used for finite-element modeling. These include:

- Euler–Bernoulli beam elements;
- plate elements; and
- shell elements.

In this book Euler–Bernoulli beam elements with the following cubic shape functions were used:

$$N_1(\xi) = \frac{1}{4}(1-\xi)^2(2+\xi) \tag{A.9}$$

$$N_2(\xi) = \frac{1}{8}(1-\xi)^3 \tag{A.10}$$

$$N_3(\xi) = \frac{1}{4}(1+\xi)^2(2-\xi) \tag{A.11}$$

$$N_4(\xi) = \frac{1}{8}(1+\xi)^3 \tag{A.12}$$

From Equations A.9 to A.12 the following characteristics may be noted:

- $N_1(-1) = 1$ and $N_1(1) = 0$;
- $N_2(-1) = 1$ and $N_2(1) = 0$;
- $N_3(-1) = 0$ and $N_3(1) = 1$; and
- $N_4(-1) = 0$ and $N_4(1) = 1$.

A.3 Estimation of Mass and Stiffness Matrices

The general formulation for the calculation of the mass and stiffness matrices can be presented in the following form (Friswell and Mottershead, 1995):

$$[M] = \int_{-1}^{1}\int_{-1}^{1}\int_{-1}^{1}[N]^T \rho[N]\det([J])d\xi_1 d\xi_2 d\xi_3 \tag{A.13}$$

$$[K] = \int_{-1}^{1}\int_{-1}^{1}\int_{-1}^{1}[B]^T [D][B]\det([J])d\xi_1 d\xi_2 d\xi_3 \tag{A.14}$$

Here, ρ is the density, superscript T stands for the transpose, det stands the determinant, [D] is the elasticity matrix, [N] is the shape function matrix, [J] is the Jacobian matrix and [B] is the matrix of the derivatives of the shape function matrix, which may be written as follows:

$$[B]=\left[\frac{\partial N_j}{\partial x},\frac{\partial N_j}{\partial x};\frac{\partial N_j}{\partial x}\right], j=1,\ldots,n \tag{A.15}$$

The Jacobean matrix describes the relationship between the local and the global coordinates. This relationship may be written mathematically as follows:

$$dxdydz=\det([J])d\xi_1 d\xi_2 d\xi_3 \tag{A.16}$$

and

$$[J]=\begin{bmatrix} \frac{\partial x}{\partial \xi_1} & \frac{\partial y}{\partial \xi_1} & \frac{\partial z}{\partial \xi_1} \\ \frac{\partial x}{\partial \xi_2} & \frac{\partial y}{\partial \xi_2} & \frac{\partial z}{\partial \xi_2} \\ \frac{\partial x}{\partial \xi_3} & \frac{\partial y}{\partial \xi_3} & \frac{\partial z}{\partial \xi_3} \end{bmatrix} \tag{A.17}$$

In the case of a one-dimensional Euler–Bernoulli beam element, Equations A.13 and A.14 can be simplified into (Friswell and Mottershead, 1995):

$$[M]=\rho A\int_{-1}^{1}[N]^T[N]\frac{\partial x}{\partial \xi_1}d\xi_1 \tag{A.18}$$

and

$$[K]=EI\int_{-1}^{1}[B]^T[B]\frac{\partial x}{\partial \xi_1}d\xi_1 \tag{A.19}$$

Here, $[B]$ represents the second derivatives of the shape function, E is the modulus of elasticity and I is the moment of inertia. For example, an element stiffness matrix for a Euler–Bernoulli element may be written as shown in Equation A.20 (Bathe, 1982).

The aggregation of mass and stiffness matrices of each element is what makes the total finite element model. The total mass and stiffness matrices are usually sparse if the orientation and the sorting of the nodes in the total system are handled well.

$$[k_e] = EI\begin{bmatrix} \frac{12}{h^3} & -\frac{6}{h^2} & -\frac{12}{h^3} & -\frac{6}{h^2} \\ -\frac{6}{h^2} & \frac{4}{h} & \frac{6}{h^2} & \frac{2}{h} \\ -\frac{12}{h^3} & \frac{6}{h^2} & \frac{12}{h^3} & \frac{6}{h^2} \\ -\frac{6}{h^2} & \frac{2}{h} & \frac{6}{h^2} & \frac{4}{h} \end{bmatrix} \quad \text{(A.20)}$$

A.4 Multi-degree-of-freedom Mass–Spring System

The mathematical representation of a finite-element model can be described in terms of the discrete mass and stiffness matrices as follows:

$$[M]\{\ddot{x}\} + [K]\{x\} = \{f(t)\} \quad \text{(A.21)}$$

Here, $[M]$ is the mass matrix, $[K]$ is the stiffness matrix, $\{f(t)\}$ is the force vector, $\{x\}$ is the displacement vector and $\{\ddot{x}\}$ is the acceleration vector. If it is assumed that the displacement response is harmonic then this can be written mathematically as follows:

$$\{x(t)\} = \{x(\omega)\}e^{i\omega t} \quad \text{(A.22)}$$

Then, the eigenvalue problem can be written as:

$$[K]\{\phi\}_j = \lambda[M]\{\phi\}_j,\ j = 1,\ldots,n \quad \text{(A.23)}$$

Here, $\lambda_j = \omega_j^2$ is the *j*th eigenvalue and $\{\phi\}_j$ is the corresponding eigenvector. For normalized eigenvectors, the following relationships can be written:

$$[\phi]^T[M][\phi] = [diag(\lambda_i)] = [\Lambda] \quad \text{(A.24)}$$

and

$$[\phi]^T[K][\phi] = [I] \quad \text{(A.25)}$$

A.5 Damping

The modeling of damping is one of the most difficult problems in applied mechanics. Fortunately, damping is adequately low in many structures and as such can be ignored in many classes of problems. The equation of motion can be written to include damping in an idealized form as follows:

$$[M]\{\ddot{x}\}+[C]\{\dot{x}\}+[K]\{x\}=\{f(t)\} \tag{A.26}$$

Here, $[C]$ is the damping matrix.

One damping model assumes that damping is proportionally related to the mass as well as the stiffness matrices and this is known as proportional damping:

$$[C]=\alpha[M]+\beta[K] \tag{A.27}$$

Here, α and β are constants. There is no physical rationale for this model but it has been found to work, particularly for lightly damped structures. The following mathematical relationship may be applied:

$$\{x\}=[\Phi]\{\nu\} \tag{A.28}$$

Here, $\{\nu\}$ is the modal participation factors vector. When Equations A.28 and A.26 are combined, the following expression is obtained:

$$\{\ddot{\nu}\}+[Z]\{\dot{\nu}\}+\Lambda\{\nu\}=[\Phi]^T\{f\} \tag{A.29}$$

Here,

$$[Z]=diag(2\zeta_j\omega_j) \text{ where } \zeta_j=\frac{\alpha}{2\omega_j}+\frac{\beta\omega_j}{2}.$$

This effectively decouples the partial Equation A.26. This, therefore, implies that for viscous damping, for the ith mode the following expression can be written:

$$\{\ddot{\nu}\}_j+2\zeta_j\omega_j\{\dot{\nu}\}_j+\omega_j^2\{\nu\}_j=\{\phi\}_j^T\{f\} \tag{A.30}$$

For hysteresis damping the following equation of motion can be mathematically formulated:

$$\{\ddot{\nu}\}+\Lambda([I]+i[N])\{\nu\}=[\Phi]^T\{f\}e^{i\omega t} \tag{A.31}$$

Here,

$$[N] = diag(\eta_j) \text{ where } \eta_j = \frac{\alpha}{\omega_j^2} + \beta .$$

Consequently, the uncoupled equation can be written as:

$$\{\ddot{v}\}_j + \omega_j^2 (1 + i\eta_j)\{v\}_j = \{\phi\}_j^T \{f\} e^{i\omega t} \quad \text{(A.32)}$$

A.6 Eigenvalues and Eigenvectors

When a viscous damping model is applied in a proportional damping model then the following eigenvalues are obtained:

$$\lambda_j, \bar{\lambda}_j = -\zeta_j \omega_j \pm i\omega_j \sqrt{1 - \zeta_j^2} \quad \text{(A.31)}$$

Here,

$\bar{\lambda}_j$ is the complex conjugate.

Alternatively, for hysteretic damping the following eigenvalues can be identified:

$$\lambda_j = \pm\omega_j \sqrt{1 + i\eta_j} \quad \text{(A.32)}$$

There are many numerical models that can be used for the calculations of the eigenvalues and eigenvectors and some of these include (Arnoldi, 1951; Saad, 1992) the

- Arnoldi iterative model;
- Gram–Schmidt process; and
- Lanczos algorithm.

In the subspace iteration method, p linearly independent vectors are iterated at the same time. At the end of the *n*th step, if $[\Phi]_n$ consists of the estimated values of the first *p* eigenvectors then a further *p* eigenvectors can be estimated from:

$$[K][\hat{\Phi}]_{n+1} = [M][\Phi]_n \quad \text{(A.33)}$$

$[\hat{\Phi}]_{n+1}$ represents the subspace through which $[K]$ and $[M]$ can be projected such that:

$$[K]_{n+1} = [\hat{\Phi}]_{n+1}^T [K][\hat{\Phi}]_{n+1} \quad \text{(A.34)}$$

$$[M]_{n+1} = [\hat{\Phi}]_{n+1}^T [M][\hat{\Phi}]_{n+1} \tag{A.35}$$

Therefore, the eigenvalues and eigenvectors projected can be written as:

$$[K]_{n+1}[Q]_{n+1} = [M]_{n+1}[Q]_{n+1}[\Lambda]_{n+1} \tag{A.36}$$

This gives an improved approximation for the p eigenvectors that can be written as:

$$[\Phi]_{n+1} = [\hat{\Phi}]_{n+1}^T [Q]_{n+1} \tag{A.37}$$

If the initial subspace is not orthogonal to one of the desired eigenvectors, then as $n \to \infty$

$$[\Lambda]_{n+1} \to [\Lambda] \quad [\Phi]_{n+1} \to [\Phi] \tag{A.38}$$

A.7 Frequency-response Functions

The frequency-response functions (FRFs) are the ratio between the Fourier transform of the response and the Fourier transform of the excitation force. The Fourier transform of a signal $f(t)$ can be calculated as follows:

$$F(\omega) = \int_{-\infty}^{+\infty} f(t) e^{-i\omega t} dt \tag{A.39}$$

The FRFs can be represented in the *inertance* or the *receptance* form. The *inertance* is the ratio between the Fourier transform of the acceleration response and the Fourier transform of the excitation force, while the *receptance* is the ratio between the Fourier transform of the displacement response and the Fourier transform of the excitation force. To calculate the FRFs, auto-spectral densities and cross-spectral densities are normally used. The following relationships can be established between the spectral densities and the Fourier transform of the signals (Ewins, 1984; Newland, 1985):

$$S_{xx} = X(\omega)\overline{X}(\omega) \tag{A.40}$$

$$S_{xf} = X(\omega)\overline{F}(\omega) \tag{A.41}$$

$$S_{ff} = F(\omega)\overline{F}(\omega) \tag{A.42}$$

$$S_{fx} = F(\omega)\overline{X}(\omega) = S_{xf} \tag{A.43}$$

Here, S_{xx} and S_{ff} are the auto-correlations of the response and excitation force respectively, while S_{xf} and S_{fs} are the cross-correlations, whereas $X(\omega)$ and $F(\omega)$ are the Fourier transform of the response and the excitation, respectively. Therefore, the receptance may be estimated as (Newland, 1985):

$$H_1 = \alpha(\omega) = \frac{S_{xf}(\omega)}{S_{ff}(\omega)} \tag{A.44}$$

$$H_2 = \alpha(\omega) = \frac{S_{xx}(\omega)}{S_{fx}(\omega)} \tag{A.45}$$

The estimators are the same if there is no averaging. The estimator H_1 minimizes the error in the FRFs due to the error in the response signal, while H_2 minimizes the error in the FRFs due to noise in the excitation signals.

Alternatively, from the estimated mode shapes and natural frequencies, the frequency-response functions can be estimated by using the modal summation equation. For a viscously damped system, the frequency-response function, H, corresponding to the excitation at position p and response measurements at position q can be mathematically written as (Ewins, 1984):

$$H_{pq}(\omega) = \sum_{j=1}^{r} \frac{\{\phi\}_{j,p}\{\phi\}_{j,q}}{\omega_j^2 - \omega^2 + 2i\zeta_j\omega_j\omega} \tag{A.46}$$

Alternatively, for hysteretic damping, the frequency-response functions can be written as (Ewins, 1984):

$$H_{pq}(\omega) = \sum_{j=1}^{r} \frac{\{\phi\}_{j,p}\{\phi\}_{j,q}}{\omega_j^2 - \omega^2 + i\eta_j\omega_j^2} \tag{A.47}$$

Here, r is the number of modes under consideration.

A.8 Modal Property Extraction

Once the frequency-response functions have been estimated, the mode shapes may be estimated by minimizing the distance between the FRFs estimated from measurements and the FRFs estimated from the model as follows:

$$J = \sum_{q=1}^{Q} \sum_{p=1}^{P} \left| H_{pq}^{model}(\omega) - H_{pq}^{measured}(\omega) \right| \tag{A.48}$$

Assuming that the system in question is viscously damped and therefore using Equation A.46, Equation A.48 may be rewritten as follows:

$$J = \sum_{q=1}^{Q} \sum_{p=1}^{P} \left| \sum_{j=1}^{r} \frac{\{\phi\}_{j,p} \{\phi\}_{j,q}}{\omega_j^2 - \omega^2 + 2i\varsigma_j \omega_j \omega} - H_{pq}^{measured}(\omega) \right| \tag{A.48}$$

By minimizing Equation A.48, the modal properties are thus estimated from the measured FRFs.

References

Arnoldi WE (1951) The Principle of Minimized Iterations in the Solution of the Matrix Eigenvalue Problem. Quarterly of Applied Mathematics, 9:17–29

Bathe K-J (1982) Finite Element Procedures in Engineering Analysis. Prentice-Hall, Englewood Cliffs

Ewins DJ (1984) Modal Testing: Theory and Practice. John Wiley

Friswell MI, Mottershead JE (1995) Finite Element Model Updating in Structural Dynamics. Kluwer Academic Publishers Group, Norwell

Newland DE (1985) An Introduction to Random Vibrations and Spectral Analysis. Longman Group.

Saad Y (1992) Numerical Methods for Large Eigenvalue Problems. Manchester University Press, Manchester

Zienkiewicz OC (1986) The Finite Element Method. McGraw-Hill, New York

Appendix B

Introduction to Vibration Analysis

B.1 Introduction

This appendix presents some practical basics of vibration analysis. Issues that are discussed include excitation and response measurements, amplifiers, filters, data-logging systems and signal processing. More details can be found in Ewins (1995).

B.2 Excitation and Response Measurements

Vibration analysis is performed by exciting a structure and then measuring the responses. A number of steps and instrumentats are needed for vibration analysis. These include (Friswell and Motterhead, 1995):

- the mounting system;
- the exciting system;
- measuring devices to measure the excitation and responses; and
- the device for recording and analyzing the data.

The most popular way the structure to be analyzed is mounted is to create a free-free environment. This can be achieved by using light elastic rubber bands. *Free-free* testing means that the structure in question is not linked to the ground. In this situation, the first modes are the rigid-body modes, which occur at 0 Hz.

There are a number of ways of exciting a structure. These include the use of a modal hammer or the use of an electromagnetic or electro-hydraulic shaker. The advantage of using a shaker to excite a structure over the use of a hammer is that the shaker can exert more energy than a hammer.

As an example, Marwala (2001) used a modal hammer to excite the cylinders. The modal hammer consisted of three main components:

- a handle;
- a force transducer; and
- a hammer tip.

The impact force of the hammer depends on the mass of the hammer and the velocity of the impact. When a modal hammer was used to hit the structure, the operator usually controlled the velocity of impact rather than the force itself. The most appropriate way of adjusting the force of the impact was to adjust the mass of the hammer. The frequency range excited by the hammer depends on the mass of the hammer tip and its stiffness.

The hammer tip had a resonance frequency above which it is difficult to deliver energy into the structure. This resonance frequency may be calculated (contact stiffness/mass of the tip). One force transducer that was used to measure the excitation force was a PCB A218.

The response from the excitation was measured using an accelerometer. An accelerometer used in the past was a DJB piezoelectric accelerometer.

B.3 Amplifiers

Signals from devices such as the impulse hammer and the accelerometer give small charges. As a result, the signals needed to be amplified by using a charge amplifier. For example, in the experiment by Marwala (2001) the acceleration signal was amplified by using a charge amplifier with a sensitivity of 14 mV/pC, and the impulse signal was amplified by using a charge amplifier with a sensitivity of 2.0 mV/pC. These amplifiers had a frequency range of 0.44–10 kHz.

B.4 Filter

One problem associated with modal testing is the problem of aliasing. When a vibration signal is measured, it had to be converted from analog into digital form, so it was sampled by an analog-to-digital (A/D) converter. This required that a sampling frequency was chosen. If the signal had significant variation over a short time then the sampling frequency had to be high enough to provide an accurate approximation of the signal that was being sampled. Significant variation of a signal over a short period of time usually indicates that high-frequency components were present in the signal. If the sampling frequency was not high enough, then high-frequency components were not sampled correctly, resulting in the problem called *aliasing*, which is a phenomenon that arises as a result of discretizing a signal that was originally continuous. The discretization process may misinterpret high-frequency components of the signal if the sampling rate was too slow, and this may have resulted in high-frequency components appearing as low-frequency components.

During data acquisition, the data were sampled at a rate that was at least twice the signal frequency to prevent the problem of aliasing. This rate was to satisfy the Nyquist–Shannon theorem (Ewins, 1995). In addition, an anti-aliasing filter may be used before the analog signal is converted into digital format to avoid the aliasing problem. An anti-aliasing filter is a low-pass filter that only allows low frequencies to pass through. This filter essentially cuts off frequencies higher than about half of the sampling frequency. As an example, in the study by Marwala (2001), the impulse and the response signals were filtered using the VBF/3 Kemo filter with a gain of 1 and a cut-off frequency of 5 kHz.

B.5 Data-logging System

The National Instruments DAQCard 1200 with 12-bit over ±5 V analog–digital conversion was used to log the impulse force and the acceleration response (Marwala, 2001). A Visual Basic program running on a Daytek desktop computer that controls the DAQCard was used to start the data logging, set the sampling frequencies, check the sample saturation and save the data.

B.6 Signal Processing

When all the measurements were taken, the next step was to process the data. The data were then processed using the fast Fourier transform (FFT) to calculate the frequency-response functions, as described in Appendix A (Marwala, 1997).

References

Ewins DJ (1995) Modal Testing: Theory and Practice. Research Studies Press, Letchworth

Friswell MI, Mottershead JE (1995) Finite Element Model Updating in Structural Dynamics. Kluwer Academic Publishers Group, Norwell

Marwala T (1997) A Multiple Criterion Updating Method for Damage Detection on Structures. Masters Thesis, University of Pretoria

Marwala T (2001) Fault Identification Using Neural Networks and Vibration Data. Doctoral Thesis, University of Cambridge

Biography

Tshilidzi Marwala, was born on 28 July 1971 in Venda (Limpopo, South Africa), and is the Dean of Engineering at the University of Johannesburg. He was previously an Adhominem Professor of Electrical Engineering, the Carl and Emily Fuchs Chair of Systems and Control Engineering, as well as the DST/NRF South Africa Research Chair of Systems Engineering at the University of the Witwatersrand.

He is a Professor Extraordinaire at the University of Pretoria and is on boards of EOH (Pty) Ltd and City Power Johannesburg (Pty) Ltd. He is a Fellow of the following institutions: Royal Society of Arts, the Council for Scientific and Industrial Research, South African Academy of Engineering, South African Academy of Science and Royal Statistical Society. He is a senior member of both the IEEE and the ACM. He is a trustee of the Bradlow Foundation as well as the Carl and Emily Fuchs Foundation.

He is the youngest recipient of the Order of Mapungubwe and was awarded the President Award by the National Research Foundation. In 2009 he won the TWAS-AAS-Microsoft Award.

He holds a Bachelor of Science in Mechanical Engineering (Magna Cum Laude) from Case Western Reserve University, a Master of Engineering from the University of Pretoria, a Ph.D. in Engineering from St John's College, University of Cambridge and completed a Program for Leadership Development at Harvard Business School. He was a post-doctoral research associate at the Imperial College of Science, Technology and Medicine and was a visiting fellow at Harvard University, Wolfson College (Cambridge) and has been elected a Visiting Scholar at the University of California, Berkeley.

His research interests include the application of computational intelligence to engineering, computer science, finance, social science and medicine. He has supervised 40 masters and Ph.D. students, published over 200 refereed papers, holds 3 patents and authored 2 books.

Index

MIX
Papier aus verantwortungsvollen Quellen
Paper from responsible sources
FSC® C105338

If you have any concerns about our products,
you can contact us on
ProductSafety@springernature.com

In case Publisher is established outside the EU,
the EU authorized representative is:
Springer Nature Customer Service Center GmbH
Europaplatz 3, 69115 Heidelberg, Germany

Printed by Libri Plureos GmbH
in Hamburg, Germany